Active Experiences for Active Children

Active Experiences for Active Children

SCIENCE

THIRD EDITION

Carol Seefeldt

Professor Emeritus, Late, of the Institute for Child Study
University of Maryland, College Park

Alice Galper

Educational Consultant

Ithel Jones

Florida State University

Boston Columbus Indianapolis New York San Francisco Upper Saddle River
Amsterdam Cape Town Dubai London Madrid Milan Munich Paris Montreal Toronto
Delhi Mexico City Sao Paulo Sydney Hong Kong Seoul Singapore Taipei Tokyo

Vice President and Editor in Chief: Jeffery W. Johnston
Senior Acquisitions Editor: Julie Peters
Vice President, Director of Marketing: Margaret Waples
Senior Marketing Manager: Christopher Barry
Senior Managing Editor: Pamela D. Bennett
Senior Project Manager: Mary M. Irvin
Production Manager: Susan Hannahs
Senior Art Director: Jayne Conte
Cover Designer: Karen Salzbach
Cover Art: Fotosearch
Photo Researcher: Lori Whitley
Full-service Project Manager: Vishal Gaudhar/Aptara®, Inc.
Composition: Aptara®, Inc.
Text Printer/Bindery: Edwards Brothers Malloy
Cover Printer: Edwards Brothers Malloy
Text Font: Times

Every effort has been made to provide accurate and current Internet information in this book. However, the Internet and information posted on it are constantly changing, so it is inevitable that some of the Internet addresses listed in this textbook will change.

Photo Credits: Shutterstock, pp. 2, 109; Alice Galper, pp. 4, 9, 15, 27, 36, 46, 48, 62, 70, 84, 88, 145, 158; David Mager/Pearson Learning Photo Studio, p. 11; Ithel Jones, pp. 16, 26, 54, 72, 164, 170, 171; Timothy P. Dingman/PH College, p. 23; Bob Nichols/USDA Natural Resources Conservation Services, p. 31; Lynn Betts/USDA Natural Resources Conservation Services, p. 39; Anthony Magnacca/Merrill, pp. 43, 124; Anne Vega/Merrill, pp. 59, 82, 129; Laimute Druskis/PH College, p. 65; Fotolia, LLC – Royalty Free, p. 98; George Doyle/Getty Images, Inc. – Stockbyte, p. 103; Bruce Johnson/Merrill, p. 106; Todd Yarrington/Merrill, pp. 114, 137; Lloyd Lemmerman/Merrill, p. 119; KS Studios/Merrill, p. 132; Ben Chandler/Merrill, p. 144; Merrill, p. 151; Laura Bolesta/Merrill, p. 155

Library of Congress Cataloging-in-Publication Data

Seefeldt, Carol.
Active experiences for active children. Science / Carol Seefeldt, Alice Galper, Ithel Jones. — 3rd ed.
p. cm.
ISBN-13: 978-0-13-265955-0
ISBN-10: 0-13-265955-7
1. Science—Study and teaching (Early childhood)—Activity programs. I. Galper, Alice. II. Jones, Ithel. III. Title.
LB1139.5.S35S44 2012
372.7'044—dc22

2011015589

10 9 8 7 6 5 4 3 2

ISBN 10: 0-13-265955-7
ISBN 13: 978-0-13-265955-0

Preface

Young children learn science by doing science. In the early childhood classroom, science involves actively engaging children in meaningful hands-on experiences. Teachers should plan science activities that build on children's interests and provide opportunities for open-ended inquiries. This book will help all teachers in planning and implementing meaningful thematic science investigations.

Grounded in John Dewey's philosophy that all genuine education comes through experience but that not all experiences are equally educative, *Active Experiences for Active Children: Science* answers teachers' questions about experiences and activities that will nurture young children's love and enthusiasm for science. In this book, teachers will learn how to plan and implement meaningful thematic investigations that truly educate young children—not just keep them busy. Teachers are guided in planning and implementing a curriculum that will actively engage children so that they ask questions, investigate, pose challenges, solve problems, make discoveries, and enhance and extend their understanding of scientific concepts.

NEW FEATURES IN THIS EDITION

This revised edition has a number of additions that will meet the needs of preservice as well as inservice teachers.

The content has been expanded in this third edition to include references to the most recent reports from the National Academies. The attractive new photos are complemented by updated and expanded lists of books for children and adults. In addition, each of the active experiences chapters in Part Two includes new and exciting science activities. Completing the new edition is a brand-new Chapter 14 on engaging children with the natural world.

New to the third edition:

- Content in Part One chapters has been updated, including references to recent research and science education reports. The content has been revised to reflect what we now know about how young children learn science.
- The new Chapter 14, "Engaging Children with the Natural World: Environmental Education," explains why children need to get outside and engage with nature. This chapter includes resources and activities to help teachers plan activities that will engage children with the natural world.
- The lists of books for children and adults in each of the content areas have been updated and expanded.
- The content in the chapters in Part Two has been extended. Each chapter includes new and exciting activities for teachers to use with their students as well as additional and updated resources, including websites for exploration.

- The tear-out sheets at the ends of Chapters 6 through 14 have been revised and updated.
- New sample letters to families have been included in Chapters 6 through 13.
- New photographs and images help create an attractive and user-friendly text.

MEANINGFUL EXPERIENCES

Active Experiences for Active Children: Science consists of clear, concise, usable guides for planning meaningful learning experiences in science for children in child-care settings, preschool programs, Head Start and other federally funded programs, and kindergarten. Primary-grade children should be engaged in active experiential learning as well, and each experience is extended to the early primary grades (grades 1–3).

The experiences in this book are meaningful because they

- Are grounded in children's interests and needs in their here-and-now world.
- Have integrity in terms of content keyed to science standards.
- Involve children in group work, investigations, or projects based on inquiry learning.
- Have continuity: One experience builds on another, forming a complete, coherent, integrated learning curriculum for young children and connecting the early childhood setting to children's homes and communities.
- Provide time and opportunity for children to think and reflect on their experiences.
- Provide the teacher with the opportunity to document and assess children's learning.

AUDIENCE

This book was written primarily for early childhood teachers at the preservice as well as the inservice levels of professional development. The text is also for those who work with teachers at all levels, such as teacher educators and administrators. It is suitable as a core or supplemental text in community college and four-year college or university early childhood courses.

The book is titled *Active Experiences* because its approach develops the background knowledge and skills that teachers need to effectively plan and implement meaningful, hands-on learning experiences. A wealth of open-ended inquiry activities are suggested for young children to do in class to further their understanding of science concepts. Professionals working in the field of early childhood education will find that *Active Experiences for Active Children: Science* supports their growth and understanding of how to put theory into practice. When used in conjunction with the other *Active Experiences* books (*Social Studies* and *Mathematics*), it provides a complete curriculum for the early childhood classroom.

ORGANIZATION

The text is organized into two parts. Part One addresses the theoretical background of science learning and describes how to plan and implement experiential learning. Chapters 1 through 5 provide a context or background for teachers to learn about the reasons for and theoretical structures that underlie the practice of science teaching in early childhood education. The first chapter includes cognitive theory, teaching strategies for science learning based on the standards, an introduction to general themes from the standards, and a section on observation, documentation, and assessment. Chapters 2 and 3 cover indoor and outdoor environments for learning, especially in the area of science. Here, emphasis is given to the teacher's role. In Chapter 4, we discuss the importance of building connections to home and the community for science learning. There are many suggestions for involving families in the process. Chapter 5 reviews research and theory and discusses science content, methodology, and teaching strategies.

In Part Two, we address the "how" of active experiences in science by providing tried and tested examples. Each of the guides to active experiences chapters (Chapters 6 through 14) is organized around themes that we have found particularly useful in early childhood classrooms. Furthermore, the suggested experiences are based on content suggested by the *Benchmarks for Science Literacy* and the *National Science Education Standards*. Teachers learn how to organize children's experiences around the identified themes. These chapters include sections for the teacher and for the children.

The "For the Teacher" sections begin by identifying concepts that are key to learning science. Goals and objectives are stated. These sections discuss concrete ideas for connecting a child's home and family to school and describe how to document and assess children's science learning. The "For the Children" sections consist of ideas for implementing the identified goals and objectives through thematic, integrated, and continuous experiences. Each experience chapter has examples of letters from teachers to communicate with families and forms that can be used to assess and document children's science learning.

USE OF THIS TEXT

We hope this book will serve as a catalyst for teachers to begin using active science experiences in their classrooms. The book presents an argument that active science investigations in early childhood represent the best practice as defined by national organizations such as the National Association for the Education of Young Children (NAEYC). Grounded in Dewey's philosophy, we see active science learning as a developmentally appropriate approach. At the same time, we demonstrate how using an active approach for science teaching and learning can support high academic standards. By using the examples highlighted in this text, teachers can nurture children's interests, motivation, and love of science. Teachers with limited background in teaching science can draw on a wide range of resources—including those in this book—to provide an enriched science curriculum.

AUTHORS

The expertise and background of the three authors is another important feature of this text. Together, the three authors bring a unique perspective to the book. Drawing on their years of experiences working in Head Start, child care, public schools, and other early childhood settings, they bring an intimate knowledge of practice to the text. As teacher educators, they know what preservice teachers want and need in a science text. Finally, as researchers, they are familiar with the most recent theory and research in the field of early childhood science, and this expertise is represented in this new edition.

ACKNOWLEDGMENTS

This revised and expanded edition is dedicated to the memory of Carol Seefeldt, who passed away in January, 2005. Her vision and conceptual insights shaped the *Active Experiences* series of books, including this text.

The new edition would not have been possible without the expertise, support, dedication, energy, and time of several people. The editors at Pearson are acknowledged, especially Julie Peters and Mary Irvin. Both worked tirelessly and patiently to provide guidance, support, and careful editing. We are also grateful to the reviewers whose comments and suggestions helped make this new edition a better text for teachers and students: Alan Bates, Illinois State University; Gloria Boutte, University of South Carolina; Gail Feigenbaum, Northern Essex Community College; Mary Hanrahan, Northern Virginia Community College; and Karen Houser, Southern Utah University.

As teacher educators, we've worked to provide hundreds of preservice and inservice teachers with the knowledge, skills, and dispositions to successfully teach young children. Each of these individuals has helped shape our ideas about science teaching and learning in early childhood education. The preservice teachers in particular have allowed us to field-test our ideas, and they have collaborated with us to design exciting new projects for your students. We are extremely grateful to each and every one of the teachers with whom we have worked.

Ithel Jones

Brief Contents 

PART ONE
THEORY OF ACTIVE EXPERIENCES 1

PART TWO
GUIDES TO ACTIVE EXPERIENCES 53

CONTENTS

PART ONE
THEORY OF ACTIVE EXPERIENCES 1

PART ONE

Theory of Active Experiences

1

Experiences and Science in Early Childhood: Theory into Practice

All genuine education comes about through experience . . . but not all experiences are genuinely or equally educative.

John Dewey, 1938, p. 13

Active Experiences for Active Children: Science guides teachers of 3- to 5-year-old children in planning and implementing meaningful learning experiences and skills in the sciences for children in child-care settings, nursery schools, Head Start classrooms, and kindergartens. Some themes are extended and expanded with ideas for children in the early primary grades. The book is based on the premise that activities are simply that—isolated, one-shot occurrences. They begin and end quickly. They give children something to do but not something to learn.

Experiences continue. They may last a couple of hours or a day, but they usually continue over weeks or even months, as in the Project Approach (Helm & Katz, 2001). A recent visit to the Center for Young Children at the University of Maryland confirms that children ages 3 to 5 years are fully capable of sustained interest in a topic over a long period of time. The children studied gardens, rabbits, and babies in the areas of science. Each experience contained multiple activities. For example, to begin the study of gardens and gardening, the children first explored the existing classroom garden, which hadn't been tended since the previous spring. They then walked through the garden and made observational sketches to document the condition of the plants. A web was constructed with children's ideas about what they knew about gardens. Another web highlighted what could be done to improve the garden. Children researched and selected plants that would do well in various soil conditions. The garden was planted. Then, a mural was constructed to depict the plan. Finally, a party was held to introduce parents and the other classrooms to the garden. The children were divided into committees to plan and execute the party. This example demonstrates that unlike isolated activities, experiences are filled with integrated learning. This book provides in-depth science experiences for young children based on the very latest research and publications in the field of science education.

In the last 10 years, several important documents have been written by science educators that are consistent with the guidelines proposed by the National Association for the Education of Young Children (NAEYC) Early Learning Standards (2002), as well as the *National Science Education Standards* (National Research Council [NRC], 1996); *Taking Science to School: Learning and Teaching Science in Grades K–8* (Duschl, Schweingruber, & Shouse, 2007); *Ready, Set, Science!* (Michaels, Shouse, & Schweingruber, 2007); *Surrounded by Science: Learning Science in Informal Environments* (Fenichel & Schweingruber, 2010); and *Inquiry and the National Science Education Standards* (Committee on Development of an Addendum to the National Science Education Standards on Scientific Inquiry, 2000). In addition, several conferences were held to join leaders in science education with early childhood experts, including the Workshop on Mathematical and Scientific Development in Early Childhood (2005). *Science & Children,* the journal published by the National Science Teachers Association (NSTA), now includes a section entitled "The Early Years: Resources and Conversation on Pre-K to 2 Science." Teachers can read more and enter discussion online at http://nstacommunities.org/blog/category/earlyyears/.

Children begin study of gardens and gardening.

Several strong themes emerged from these efforts, including the following:

- The introduction of children to the essential experiences of science inquiry and explorations must begin at an early age.
- Science learning should be embedded in children's work and play.
- Science learning builds on children's prior experiences, backgrounds, and early theories.
- The child's prior knowledge should be assessed before teachers provide new experiences.
- Teaching preschool-level and primary-level science is an active enterprise. Rather than the memorization of facts, young children should be engaged in hands-on inquiry learning. The emphasis should be on gaining experience with natural and social phenomena and on enjoying their explorations. Fundamental concepts and skills develop as children have the chance to ask questions, conduct investigations, collect data, and apply their problem-solving skills to new situations.
- Young children should reflect on and represent their experiences and share their ideas with others.
- There must be an emphasis on group as well as individual teaching approaches. Much of science learning is a cooperative endeavor.

The documents emphasize content standards in each area of the science curriculum. The content standards are arranged by age. Building on the science standards, *Ready, Set, Science!* (Michaels et al., 2007) defines and describes four strands of science learning. These strands offer a new perspective on what is learned during science and highlight that children's knowledge is not static. The four strands are

1. Understanding scientific explorations
2. Generating scientific evidence
3. Reflecting on scientific knowledge
4. Participating productively in science

In our increasingly technological society, it is important not only that young children are introduced to important concepts in each of the content areas of the sciences, but also that they explore these concepts using the multifaceted techniques of scientific inquiry. Children are natural scientists. They enjoy making observations, asking questions, planning investigations with the help of the teacher, and using tools to gather, analyze, and interpret the data they collect. They propose answers and communicate the results to others in a variety of ways.

The content suggested by these documents and the experiences that follow in this book have integrity and meaning. For example, what content are children learning when they are asked to make a cherry tree by pasting cotton balls that the teacher has painstakingly tinted pink on a tree trunk drawn on a large sheet of brown paper by the teacher? Are children learning that cotton balls are cherry blossoms?

What content could children learn about cherry trees? Meaningful content might include the concepts that trees bud, bloom, and produce fruit in season. By observing cherry trees throughout the year and by making and recording predictions of when they will bloom, what color the blossoms will be, and what will happen when they fall, children actually experience cherry blossoms and trees. Unlike learning about cherry trees by being told that pink cotton balls are cherry blossoms, children are being introduced to concepts about trees that are real, are meaningful, and have integrity.

This book makes clear connections between the active theme-based science experiences suggested in later chapters and in *The Benchmarks for Science Literacy* (American Association for the Advancement of Science [AAAS], 1993) and *The National Science Education Standards* (NRC, 1996) (see chapter 5 for the organizational scheme).

COGNITIVE THEORIES: THE BASIS OF SCIENCE EDUCATION FOR YOUNG CHILDREN

Our understanding of how children learn science has been influenced and guided by a number of theories. The role of active experiences in learning was identified by John Dewey and Maria Montessori early in the 20th century. The importance of activity was further highlighted by the contributions of Piaget, Vygotsky, Bruner, Erikson, Gardner, and others. These theorists share the view that learning is not a passive activity but a process that requires the active participation of learners in constructing their own knowledge. Modified by later researchers, the theories of John Dewey continue to be highly regarded by a large number of early childhood professionals for their emphasis on firsthand learning in the here-and-now world, child-initiated learning, and age-appropriate learning experiences and content.

Jean Piaget (1973) researched through clinical interviews the position that knowledge is constructed in the mind of the learner. He believed that young children think differently, in most circumstances, than do older children and adults. As discussed in chapter 4, Lev Vygotsky (1986) added significantly to Piaget's theories by postulating that the important factors in moving children to higher levels of thought are the significant and more accomplished others around them. Drawing on the work of Piaget, Vygotsky, Dewey, and others, the position taken in this book is that development and learning occur through a constructive process. According to this view, knowledge is not simply acquired from the environment or from others but is based on what the individual child brings in to the classroom. The mental patterns that the child has already constructed are modified and built upon as he or she tries to make sense of new experience.

In recent years, the commonly held view that young children are simplistic thinkers has been challenged. We now know that young children's concepts are much more sophisticated than had been assumed in the past (Gelman & Brenneman, 2004). During the early childhood years, children are developing an understanding of the world around them. Through all of their daily activities and interactions, young children engage in efforts to understand or make sense of their world. We also know that children from birth to age 5 engage in efforts to make sense of their world on many different levels, including language, social interaction, counting and quantifying, spatial reasoning, physical causality, problem solving, and categorization (Shonkoff & Phillips, 2000). Through these and other activities, young children develop ideas or concepts about their world. It is therefore important to build on young children's thinking and reasoning abilities and to provide opportunities for them to establish skills and ideas in science.

As a result of the synthesis of the previously mentioned theoretical positions and new thinking on children's acquisition of knowledge, the dominant theory upon which science teaching today is based is *constructivism,* a theory based on the idea that children play an active role in constructing new knowledge. It is a perspective that considers learning to be an active process that is influenced by the learner as well as by the teacher. For constructivists, learning is an interactive process and a consequence of how students encounter and process information based on their existing personal knowledge. "As the Standards point out, students have to construct, or build, their own knowledge in a process that is individual and social. Students have to take an active role in their *own* learning. This teaching/learning relationship is called constructivism" (Lowery, 1997, p. 7). Because children acquire knowledge at different paces and through different learning styles, it is also important for teachers to provide instruction that meets the individual needs of young children.

HOW CONCEPTS BUILD

Science concepts grow and develop in infancy. Current research on brain development emphasizes the importance of early stimulation in developing brain connections from birth. Although this book is not structured to provide experiences for infants, caregivers should understand that infants explore the world with their senses and should provide many opportunities for them to do so. If babies do not have things to look at, touch, smell, and hear, they will be deprived of the structures through which further knowledge is acquired. Toddlers need to be free to safely discover things on their own. They need things to grasp, carry, sort, push, and pull. Through the daily schedule, they will also begin to develop a sense of time.

Although their primary avenue for concept learning is exploration, preschool- and kindergarten-age children (sometimes with the help of an adult) can

- formulate questions, collect data, and develop answers;
- organize, reflect on, represent, and document their investigations;
- share and discuss ideas with others.

As children begin to collect and organize data to answer a question, new concepts and skills emerge.

In her book *How to Work with Standards in the Early Childhood Classroom,* Seefeldt (2005) posits that scientific inquiry includes observing, questioning, investigating, analyzing, reaching conclusions, and communicating the results to others. Young children do some of these things naturally. Others require the assistance of the teacher. For example, young children constantly observe, but scientific observation requires focused observation. The teacher should model focused observation continuously. Similarly, young children have many questions, but they do not always have the skills to articulate what they really want to know. Teachers can also model questions and provide the psychological safety that children need to keep asking. Seefeldt (2005) further suggests that teachers give up "show and tell" in favor of group meetings in the morning, at which children feel free to plan their investigations for the day or after work time as a forum for children to discuss the conclusions from their inquiries. At group meetings, children can dictate their ideas to document their learning.

SCIENCE PROCESS SKILLS

Children develop concepts as they engage in active experiences, exploration, and inquiry. To be successful at these activities, children must develop their ability to use process skills (Harlen, 2000), which include observing, measuring, classifying, communicating, estimating, predicting, and experimenting (Jones, Lake, & Lin, 2008). These skills can be used and refined in the context of active science investigations. They should also be developed during regular classroom activities because they are the same skills that are regularly used by children in developmentally appropriate early childhood programs.

TEACHING STRATEGIES FOR SCIENCE LEARNING

One of the primary roles of the constructivist teacher is to understand how each child is constructing his or her understanding of scientific experiences. Having identified the child's existing understanding, the teacher provides experiences that further develop this understanding. These experiences should enable children to reflect on their initial understanding and to reconstruct information in ways that are individually meaningful. The constructivist teacher should therefore adopt strategies that enable children to construct their own understanding. Some of the strategies listed here are outlined in the following section:

- Observe children in order to identify how they understand the concepts of interest.
- Provide learning experiences that are meaningful and relevant to the children.
- Provide opportunities for children to work with concrete materials, to make choices, and to explore.
- Listen and respond to children's thoughts and ideas.
- Ask open-ended questions.
- Provide opportunities for children to reflect on their experiences.
- Respond to children's interests.
- Select science activities that allow children to apply skills.
- Encourage children to problem solve and investigate.
- Provide opportunities for children to represent their ideas.
- Allow for differences in interests and learning styles.
- Encourage students to interact with each other and with the teacher.
- Document learning to evaluate the impact of scientific experiences in terms of knowledge construction.

Additionally, considerations of equity are critical in teaching early childhood science. Teachers should model acceptance and tolerance and support individual learning styles. All students are capable of full participation in the science program, and efforts should be made to assist children who are nonnative speakers and to compensate for children's inadequate experiential backgrounds.

The teacher should be an active participant in finding out the answers to science problems and may use several teaching strategies to facilitate the development of science concepts. The following strategies are essential to planning and participating in effective science learning:

- **The provision of an interesting and rich environment for the children.** The teacher serves as facilitator, observes the children's spontaneous activity, notes progress, and encourages the children in their explorations.
- **The initiation of informal learning experiences as a child is engaged in naturalistic exploration.** These experiences are usually not preplanned but arise when the teacher senses that a child is on the right track yet needs some additional cues to solve a problem. This strategy further involves taking advantage of the "teachable moment" to reinforce children's discoveries.
- **The planning of structured experiences for both small and large groups around the major content areas identified in *Taking Science to School* (Duschl et al., 2007) and the *National Science Education Standards* (National Academy of Sciences, 1996).** The teacher should identify major themes, such as "the earth"; gather resources for both adults and children; and plan experiences and experiments through which children can answer their questions about their earth.
- **The modeling of effective problem-solving techniques.** Teachers are powerful models for children. If they are enthusiastic and open about science teaching, children will adopt similar attitudes.

Each of these strategies is developmentally appropriate. They should all be used at one time or another as part of an effective science program. Additionally, children need plenty of time to engage in the process of science learning and to reflect on their findings.

TOOLS TO HELP CHILDREN STUDY SCIENCE CONCEPTS

Children are naturally curious about their world and will investigate it with enthusiasm, given the opportunity. Yet teachers need to provide children with tools to assist them in gathering and analyzing information. Many examples are found in the experience chapters to follow, but Seefeldt and Wasik (2006) suggest some basic ideas:

- Clipboards, paper, and markers for each child to take notes about what he or she sees on a trip and why. Younger children can draw the things they see.
- Yarn or hula hoops to place on the ground to circle things for children to observe.
- Viewing tubes to look through to spot various things in nature.
- Collection bins for children to categorize the things that they have collected from a nature walk.
- Inexpensive cameras that children can share to document what they see.

Finally, a digital camera is an invaluable tool for the teacher to record what children are doing in their scientific inquiries. The photographs may be displayed to document and remind children of science learning. They may also be duplicated for class books or family newsletters.

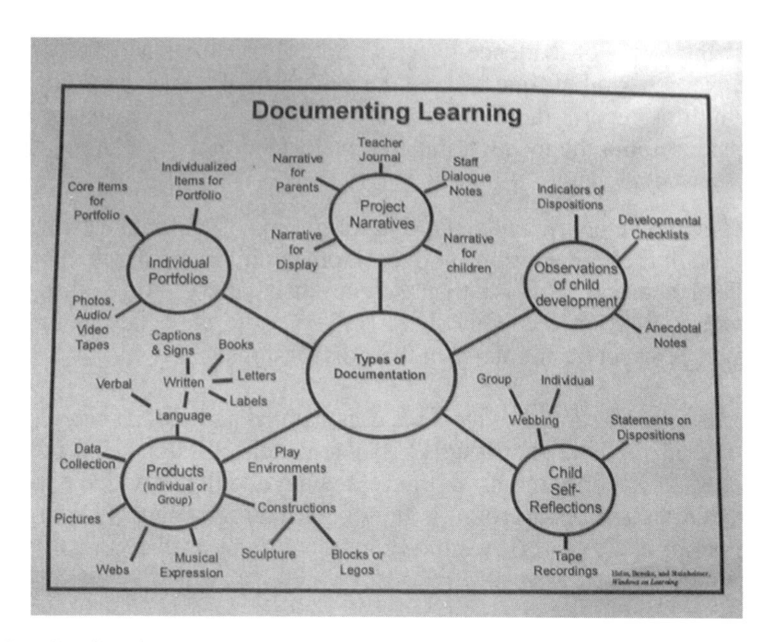

By engaging in the documentation process of collecting, describing, and interpreting evidence of young children's emerging science understanding, early childhood educators are able to provide appropriate science experiences and learning environments.

OBSERVATION AND ASSESSMENT

Teachers are often overwhelmed by the assessment process. Books in all of the content areas of the early childhood curricula suggest that teachers create portfolios of individual children's work and folders that profile the progress of the entire class. Yet teachers are puzzled about what a portfolio is and what goes into it. "A portfolio is a purposeful collection of evidence of a child's learning, collected over time, that demonstrates a child's efforts, progress, or achievement. Physically, a portfolio might be a folder, box, or drawer devoted to the collection" (McAfee, Leong, & Bodrova, 2004, p. 52). A portfolio might contain anything that the teacher or the child wants to include. More formal entries include checklists, rating scales, and anecdotal records compiled by the teacher at regular intervals. Anecdotal records capture what a child says and does during the course of a learning experience. Entries in a science journal provide excellent documentation of older children's understanding of the experience.

Less formal entries include photographs, videotapes, audiotapes, writing samples, artwork of all kinds, and lists of books read. When using photographs or items of children's work, the teacher should add a note about what the artifact means in terms of growth and development in the area of science. Portfolios should not be a secret to children or their families because they are used to interpret strengths, weaknesses, and progress over time. Children can be asked at intervals for work they would like to include.

Teachers should regularly analyze the data that they have collected. By engaging in the documentation and assessment process of collecting, describing, and interpreting evidence of young children's emerging science understandings, early childhood educators are able to provide more appropriate science-related experiences and learning environments. Jones and Courtney (2002) suggest three guiding principles in documenting early science learning:

- Collect a variety of forms of evidence. Such different forms of evidence are necessary because of the variation in the ways children convey their ideas.

- Collect the forms of evidence over a period of time. A teacher who collects evidence over a period of time can see the evolution of a concept or idea. For example, children record their observations of mealworms over time and reach conclusions about the mealworms' life cycle. (See activity 3 in the "For the Children" section of chapter 8.)
- Collect evidence on the understanding of groups of children as well as individuals. Through their interactions with each other, children enlarge their understanding of science and increase their science vocabulary. For example, the group observation chart on Tear-Out Sheet 10-2 refers to the accuracy of the group in employing correct terminology in their discussions.

Finally, the teacher should draw some conclusions and make some decisions. Are individual children meeting their goals? If not, which children need motivation and encouragement? Which children have progressed beyond the provided experiences and need greater challenges? As a group, is the class making progress toward the benchmarks described by the experts? Are modifications in the teacher's instructional practice necessary?

SUMMARY

Children learn from content-based science experiences because they are involved with their hands and their minds and are actively engaged in making sense of themselves in their world. Because experiences are embedded in children's here-and-now world, they are of interest to children. *The Benchmarks for Science Literacy* (AAAS, 1993) suggest that teachers start with questions about phenomena that are interesting and familiar to students, such as nature. This interest motivates young children to meet the challenges of a science program and become successful learners. If children are not interested in science content, they have not experienced it in a way that encourages inquiry and continuity of experience.

Theme-based experiences continue. When children leave school for the day, they know that there is something for them to continue doing when they return the next day. The fact that the experiences in this book are based on concepts that are key to the discipline not only gives them intellectual integrity but offers children continuity of content. Because the experiences are connected to children's family and community, there is a continuous thread of learning in their lives. The *Benchmarks for Science Literac*y (AAAS, 1993), *National Science Education Standards* (NRC, 1996), and this book emphasize the vital roles of families and communities in facilitating and enhancing science education in and out of the classroom. Many resources are suggested in the following chapters and in the References section.

Finally, documentation and assessment are a vital part of the science program for young children. When teachers collect children's work, keep records, and reflect upon their findings, they are able to modify old strategies and identify new strategies to meet young children's needs and the goals of the curriculum. Additionally, teachers find documentation and assessment to be a valuable way to report to parents at conferences. Actual examples of children's work are concrete for parents and essential to understanding their children's progress.

Now is an excellent time to be writing about science experiences for young children. Evidence and support are now present for early childhood educators to propose a science program that requires different content and strategies from those previously based upon memorization and rote learning. The *Benchmarks for Science Liteacy* (AAAS, 1993) and *Science Education Standards* (NRC, 1996) are consistent with developmentally appropriate practice and provide an excellent beginning for an inquiry-based science curriculum for young children.

2

Active Children—Active Indoor Environments

Scientific exploration is best cultivated through experiences that build on children's current interests and preoccupation—the skill of the practitioner is to introduce resources and ideas which will trigger these interests.

Brunton and Thornton, 2010, p. 2

Centers of learning are ideally suited for the social interaction essential to the child's construction of meaning. These centers encourage rich and challenging classroom talk between children and adult–child dialogue. This interaction in turn fosters peer-based learning and assistance from more competent adults, which enables children to reach higher conceptual levels. The range of skills that can be achieved with adult guidance and peer collaboration far exceeds what a child can learn alone or from whole-class instruction. This chapter presents general guidelines for designing environments for children—health and safety, inclusion, and beauty. More emphasis is placed on science centers, but the whole environment is considered to be a laboratory for learning.

If active children are to learn through active experiences, their environment must be carefully, thoughtfully, and deliberately arranged. Indoor spaces should be structured so that children can

- engage in meaningful firsthand learning, taking the initiative for their learning and making choices and decisions;
- work, play, and interact freely with others, both peers and adults;
- have the tools available to facilitate problem solving;
- use language (talking, listening, writing, and reading) in connection with their interactions with the physical world and socially with others;
- experience success, risk, and challenges as they gain new skills through interaction with their physical environment;
- be alone so that they can reflect on their experiences.

Beginning with the essentials—health, safety, inclusion, and beauty—teachers plan for children's meaningful learning experiences by deliberately arranging the indoor learning environment for active learning, in addition to planning ways of interacting with children that foster and promote children's learning and development.

THE ESSENTIALS: HEALTH, SAFETY, INCLUSION, AND BEAUTY

Health and Safety

Specific health and safety concerns are delineated for each of the activities in the following chapters. The following are more general concerns. The indoor environment must be set up with each child's health and safety in mind (Isbell & Isbell, 2005):

- Equipment is checked for sharp edges, loose pieces that could cause accidents, or small parts that children might swallow or stuff in their nose or ears.

- Most materials provided for the children are unbreakable, although in some instances—such as making an aquarium—glass is the best choice, and adult guidance is essential

- Equipment is disinfected by daily washing with detergent in water, rinsing with clear water, wiping or spraying with a solution of 2 tablespoons of chlorine bleach and 1 gallon of water, and sun- or air-drying.
- Use of all heat sources is well supervised, and outlets are covered when not in use.
- Tubs and pools of water are closely supervised.
- Tools such as knives and hammers are in good condition, and their use is carefully observed.

Safety is a fundamental concern in the science curriculum. Teachers must know and apply the necessary safety regulations in the storage, use, and care of the materials used by students. Locked cabinets are a must when materials are not in use. In addition, careful rules must be established for the care of any fish and small pets that reside in the classroom. Parents should be contacted about any allergies that would prevent their child from participating in a particular science project.

Space should also be available to provide children with active experiences that foster healthy physical and psychological development. Both the *National Science Education Standards* and *Benchmarks for Science Literacy* (AAAS, 1993) suggest that students become acquainted with factors that foster healthy bodies and develop an understanding of pleasant and unpleasant feelings.

Inclusion

The physical environment can be arranged in ways that enable all children to participate to the fullest extent possible in all experiences. To permit use of a wheelchair, remove physical barriers, provide wider paths, and arrange work spaces and activity units to offer shelter from intrusion or interference (Klein, Cook, & Richardson-Gibbs, 2001). More accomplished peers may model and teach the use of materials in the science center. When safety is an issue, a "buddy" may be assigned to a child who is less able. Children with special needs may profit from a multisensory approach to the teaching of science.

Reducing the amount of visual stimulation in a given area aids children who are visually impaired. Teachers have found that they can add textures or raised patterns to the walls to enable visually impaired children to locate themselves in space. Tactile experiences are also valuable for nonimpaired peers because they foster learning through touch. Others find small shelving units, with a few materials on each shelf, to be helpful.

Children with hearing impairment require more visual stimulation and less auditory distraction. Felt pads on tabletops, carpeted shelves and other work surfaces, and the clear display of all materials and equipment will be helpful (Klein et al., 2001). Teachers can actively foster children's understanding and appreciation of the senses and sensory impairments through active experiences with the five senses. (Specific activities are discussed in chapter 12.)

Gould and Sullivan (1999) do not believe that teachers need to reduce the amount and variety of learning materials within a classroom to accommodate the needs of children with physical disabilities. Instead, they suggest modifying and adapting the activities: "All children in early childhood classrooms need nurturing and stimulating learning environments, but for children with special needs, many typical experiences need to be modified to promote their development" (p. 13). Mallory (1998) suggests that small groups consisting of two to four learners provide an optimal structure for fostering cognitive development and social participation in inclusive classrooms: "Given the heterogeneity that exists in inclusive classrooms, it is logical to assume that whole group activity is not likely to be an effective means for assuring that the particular needs of individual learners will be met" (p. 228).

Beauty

Aesthetics and beauty must be considered for young children growing up in a world that increasingly prefers highways and large housing developments to green space and historic buildings. Aesthetics means being sensitive to beauty in the environment—in nature and art. "Such sensitivity is fostered not by talking about beauty but by experiencing it in a variety of forms" (Wilson, 1995, p. 4). It is important to note that beauty resides not only in the natural environment but also in vibrant city neighborhoods and small historic towns.

The child-care centers in Reggio Emilia, Italy, illustrate the wonder and beauty of environments created with aesthetics in mind. Stepping into a child-care center in Reggio, one knows immediately that the environment has been carefully arranged to simplify and order the children's world as well as surround them with beauty: "Educators in the Reggio Emilia schools have helped us become more aware of the beauty that is possible in early childhood settings" (Isbell & Exelby, 2001, p. 26).

Open rooms filled with light and air are simply and elegantly arranged. This clear, clean conceptualization of an environment is especially appropriate for active children who learn through active experiences and facilitates the child's construction of beginning science concepts. By integrating conceptual learning in various subject matter areas with artistic expression, children are encouraged to make visual or symbolic representations of their activities in a variety of media.

Everywhere you look, there is something beautiful to wonder over. Mirrors of all types are found throughout the center. Bits of mirrors and colored glass hang in front of windows to catch a sunbeam and bounce it back to the children. Long horizontal mirrors are mounted near the floor so that children can watch themselves as they build with blocks or play with others.

Plants and flowers are present throughout the center in classrooms, lunchrooms, sleeping rooms, and bathrooms. Children's artwork is mounted, framed, and displayed, serving not only to stimulate children's thoughts and to permit them to document and reflect on past experiences but also to inform others of the experiences that children are having in and out of the center. The home is intimately connected with the centers.

This emphasis on aesthetics in Reggio reflects an underlying order and beauty in the design and organization of the entire space: "Every corner of every space has an identity and a purpose, is rich in potential to engage and to communicate, and is valued and cared for by children and adults" (Caldwell, 2002, p. 5). Yet children everywhere—not just in Italy—deserve to live and learn in environments that are aesthetically pleasing and visually appealing (Seefeldt & Barbour, 1998). Bredekamp (1993) reflects that "perhaps we, in America, have set our sights too low in our vision of excellence" (p. 13). Too often in programs for young children, little attention is given to the beauty of the classroom. Commercial posters and decorations take the place of objects of beauty and child-inspired art. Seefeldt (2002) maintains that children deserve to be surrounded with beauty. Because the environment affects children's behaviors as well as their art and written work, children should see something of beauty in every place they look.

INDOOR SPACES

Integrating Spaces

Organizing indoor spaces with centers of interest permits active children to engage in active experiences. Centers of interest are areas of the room that are clearly defined with either actual dividers or suggested boundaries. They contain materials and equipment organized to promote specific types of learning. The materials are carefully arranged so that children can see the choices available and make decisions about which materials

The display of artwork allows children to document and reflect on past experiences.

they will use and how they will use the materials (Bronson, 1995). In designing these contexts for scientific learning, an effort should be made to create an environment that is ordered, interesting, and beautiful (Brunton & Thornton, 2010). Educators in Reggio Emilia emphasize the importance of "relational forms, light, color, materials, smell, sound, and microclimate" of environments for young children (Ceppi & Zini, 1998).

In order to encourage investigation and exploration, spaces should be uncluttered. Spaces should also be arranged so that they can be easily changed and adapted to accommodate children's current interests. Though some areas are defined, teachers should also work toward an integrated curriculum framework so that, for example, science becomes a part of the social studies, arts and crafts, music and movement, mathematics, and emergent literacy (for younger children) and the language arts (for older children). "When we integrate science experiences with other curriculum areas, we help children enhance their mental performance. Different features of an experience are encoded in different parts of the brain. These features are then linked together in more enduring memory systems deeper in the brain" (Harlan & Rivkin, 2004, p. 11).

Woyke (2004) presents some suggestions for integrating science into various classroom areas. For example, in the art area, one suggestion is to add natural materials such as cotton, wool, pine cones, rocks, leaves, and bark to use in prints, rubbings, and collage. In the music corner, a teacher could offer CDs featuring sound recordings of nature, such as the wind in the trees, ocean waves, and rain. Teachers could also use natural materials such as shells as musical instruments.

Brian the hamster goes home with different children over the weekend. Parents and children record their experiences with Brian in a journal.

Children's literature is an excellent way to integrate science throughout the early childhood curriculum. An important part of an effective science curriculum is a classroom library with a wide variety of science-related children's literature, both fiction and nonfiction. Teachers may read books such as *Everybody Needs a Rock* by B. Baylor (1985) to stimulate children's interest in rocks and minerals. Stories, guides, and reference books should also be placed in the various centers of interest for inquiry and stimulation. For example, as part of a theme on pets, children can

- read about various pets of fact and fiction;
- find out what various pets eat, their preferred habitat, and how to care for them;
- chart the habits of classroom pets (when they eat, sleep, play);
- compare pets;
- write about pets using science journals, posters, poetry, and stories;
- create movement activities, finger plays, and drama based on pets;
- depict pets using many art forms;
- develop a caring and sensitive attitude toward pets.

In addition, if pets are kept in the classroom, children can make and record rules for their care and handling. At the Center for Young Children at the University of Maryland, Brian the hamster goes home with various children on weekends. With the help of parents, children record their experiences with Brian in a journal that then passes to the next child. During the week, the journal resides in the literacy area to be examined by the children. The integrated curriculum is emphasized in the following chapters.

Science Areas

Taking Science to School (Duschl et al., 2007) suggests that "Students learn science by actively engaging in the practice of science. A classroom environment that provides opportunities for students to participate in scientific practices includes scientific tasks embedded in social interaction using the discourse of science and work with scientific representations and tools" (Duschl et al., 2007, p. 342). Teachers should design science areas to allow students to work safely in groups of various sizes at various tasks, to maintain their work in progress, and to display their results. Wonderful displays of the results of children's inquiry are often ignored in early childhood classrooms. They provide an opportunity for children to reflect upon their work and validate the importance of their efforts. Teachers also provide students with the opportunity to contribute their ideas about use of space and furnishings (Duschl et al., 2007).

Spaces are needed in which children can actively experiment with and explore the life, health, and physical sciences. Natural science and earth and space science experiences may be best reserved for the outdoor environment, although there are many opportunities to construct experiments and displays indoors. The center or centers should provide opportunities for children to observe, classify, compare, measure, communicate, experiment, make predictions, and reach conclusions. The teacher should provide materials that excite children's natural interest, encourage many questions, and promote active inquiry.

One teacher arranged a pitcher of water and small cups on a table next to small containers of instant coffee, tea leaves, dirt, sand, beans, sugar, and salt. She posed a question for the children: "Which things will dissolve in water and which ones will not?" A clipboard with a checklist for children to record their findings was placed beside the center. "Things That Dissolve" and "Things That Do Not Dissolve" were identified. Children were encouraged to discuss and negotiate their conclusions. The children later presented the center as part of a series of experiments for their families to try at the conclusion of a science unit. Other science learning centers could be equipped with things to weigh, measure, and balance. Magnets, compasses, prisms, magnifying glasses, and different kinds of mirrors and colored cellophane also promote children's active explorations.

Machines to take apart—clocks, pencil sharpeners, instrument-panel boards (all of which have been safety-proofed)—along with screwdrivers and wrenches fascinate children, who are curious about how things work. One group of 5-year-olds worked for days taking apart an alarm clock and recorded each step in detailed drawings that the teacher displayed to document the process and then stored for later reflection.

Chrisman (2005) suggests a "nuts and bolts" science center at which children develop skills in comparing, developing fine motor control, and classifying by size and shape. After the experience, children build graphic organizers to summarize what they have discovered. Examples include webs showing what things are held together by nuts and bolts and charts showing the different lengths of the bolts used or displaying vocabulary words that describe nuts and bolts. (See also the "Woodworking Center" section later in this chapter.)

Living things may be a part of the science area if they are well cared for and do not present safety hazards for children. It is a desirable goal for children to learn about the life cycle, yet too many premature deaths of classroom animals result from poor care and a lack of safety rules in early childhood classrooms. Plants and insects are often a better alternative. An ant farm created from a discarded large food jar, intrigues children, as does a butterfly farm or a worm garden. Sometimes teachers might invite a pet to visit the classroom. Children should consult books and formulate questions about the pet before the visit. Answers should be recorded in science journals or on chart paper.

Whatever active experiences spring from the children's interests or are planned by teachers, it is important that enough time be made available for extended investigations. Teachers make time for students to work in varied groupings—alone, in pairs, in small groups, as a whole class—and on varied tasks, such as reading, conducting experiments, reflecting, writing, and discussing. Beginning to build scientific understanding in children takes time and adjustable scheduling on a daily basis and over the school year.

Art Centers

Through the visual arts, children are able to give expression to their ideas, imaginations, feelings, and emotions. This expression is necessary for children to reflect on their experiences. Each day, children should have a choice of whether to draw, paint, model, cut and paste, or construct something. Materials are arranged on tables or shelves that are easily accessible to children. Easels, a variety of brushes, and fresh, thick paints are available every day. At other times, areas of the floor or a table or two can also be used for painting. All types of drawing materials—crayons, markers, chalk, even pencils for 5-year-olds—are stored on open shelves for children's selection (Dighe, Calomiris, & Van Zutphen, 1998). A junk box with every type of material imaginable and a sewing box for 5-year-olds equipped with threads, bits of fabric, buttons, and large blunt needles are available. There is a separate area for clay and modeling materials.

Because the visual arts give children a way to organize, reflect upon, and present their ideas or emotions, art materials are chosen that enable children to do so. For example, a group of children took a walking trip to the park to see the cherry trees blooming near their center. When they returned, they found the art center equipped with a variety of pink, lavender, and white paints and papers.

Manipulating modeling clay and observing its properties, painting with tempera paints and observing drips and how colors mix, and organizing and sorting materials for a collage help children understand science concepts and integrate them with other subject matter areas. Seefeldt (2002) emphasizes the importance of displaying children's work both to document their work for themselves and visitors and to enhance the beauty of the classroom. At the Center for Young Children, children's paintings are framed with wood, and leaf collages are laminated for extended display.

Woodworking Center

A center for woodworking gives children other opportunities to re-create their experiences. Woodworking involves such a variety of skills, tools, and materials that there are endless opportunities for using the processes of science, such as classification, comparisons, measurement, and predictions about how the finished construction will look. Through trial and error, children will also draw conclusions such as "long, fat nails will split soft wood." Children are endlessly fascinated by woodworking, but teachers must closely supervise and co-construct learning with the children.

A place for children to construct three-dimensional objects should also be included. Children can choose to work with any material. In Reggio, found objects—boxes, feathers, shells, sequins, paper, silk, and brocade scraps—are stored on open shelves in aesthetically pleasing ways, inviting children to choose the materials they will use and how they will arrange them. The design and creation of three-dimensional objects helps children master science concepts that revolve around weight and balance.

Book and Library Centers

Materials should be selected that encourage children to use written as well as spoken language, including notepads, calendars, discarded checkbooks, address books and pencils, markers, and pens for children to write with as they indulge in dramatic play. Reading materials such as newspapers, magazines, phone books, and fiction and informational books enhance play.

No two housekeeping centers will be exactly alike or equipped with the same materials because each center will contain props representing the culture and activities of the children's homes. In one kindergarten class, a diverse group of children were having a heated discussion in the housekeeping kitchen about the name of a dumpling. The teacher saw this as an opportunity to teach children how foods such as dumplings are made by families all over the world but have slightly different ingredients and different names. Through stories and cooking activities, the children explored ravioli, wontons, pierogies, and samosas.

Other areas for sociodramatic play are appropriate. When children visit their parents who work in offices or visit the school's office, an area for office play is applicable. A table framed with cardboard walls becomes an office cubicle. Telephone message pads, bookkeeping forms—anything that looks official and has space for writing—along with pencils, erasers, and markers of all kinds belong in the office. Children enjoy exploring and playing with staplers, pencil sharpeners, and other office tools, especially rubber stamps with stamp pads.

Based on children's experiences with their world, other dramatic play areas will be pertinent. For instance, if children have visited

- a post office, then a post office with envelopes, stamps, machines to weigh objects, and cubbyholes in which to sort mail would be created.
- a fast-food restaurant, then a restaurant with aprons, hats, trays, boxes for food, cups and bottles, a cash register, and pretend money would be added.
- a supermarket, then a store with a cash register, money, all types of food containers, cans, boxes, bags to pack, old cash register receipts, and other materials would be arranged so that children could reenact their visit to the supermarket, taking turns being shoppers, clerks, bakers, or shelf stockers.
- a doctor's office, then an office with a waiting room, charts of the body, stethoscopes, unbreakable thermometers, tools to test the reflexes, and white coats can be added. Bandages are a must to treat pretend patients or dolls in need of aid. Small cots may provide a temporary resting place. Writing pads and pencils should be available for dispensing prescriptions.
- a veterinarian, then an animal hospital serving stuffed animals might be created using some of the props mentioned previously.

Areas for Manipulatives

In manipulative centers, children learn a good deal about the physical sciences. Teachers arrange age-appropriate puzzles, board games, matching and bingo games, pegs and pegboards, construction sets, small plastic or wooden blocks or tiles, Tinkertoys, erector sets, large LEGOs, and other materials. All these give children needed practice in observing, ordering, discriminating, classifying, and predicting.

A few sets of regular decks of cards and materials such as large beads and buttons, nuts and bolts, washers, seashells, and other objects are important additions for sorting, counting, and categorizing. Things for children to string—beads, bottle caps with holes in them, and shoelaces—are enjoyable. Using pull toys with removable wheels to discover the difference between moving toys with wheels and without them and inventing machines using Tinkertoys and other building sets with gears, levers, or pulleys gives children the active experiences they need to build science concepts.

Block Areas

Blocks and spaces in which to build are essential. Ideally, blocks should be stored on open shelves with a place for each type of block. Storing all rectangular blocks on the same shelf, for example, fosters children's ability to classify. A complete set of wooden unit blocks is the best investment a center or program can make. If these are unaffordable, blocks can be made out of paper cartons. Science concepts are fostered as children observe, build, measure, compare, predict, begin to learn about spatial relations, and construct building plans. Chalufour and Worth (2004) provide a thorough exploration of the importance of building structures with young children in tandem with a science curriculum, including focused explorations, extension activities, and resources.

Water and Sand Areas

Children need indoor areas where they can explore the properties of both water and sand and develop concepts of the surfaces of the world in which they live. Water is easy to provide. All that is needed is a low table; a small, plastic pan; some plastic cups and containers—spoons, funnels, plastic tubing, straws; and a small amount of water. Children build science concepts through active experiences in using containers, watching water flow, making bubbles in water, and observing how water is absorbed by paper towels.

Sand—in a sand table, a plastic tub, or an old wading pool—can be readily available indoors. As with water play, children learn beginning concepts through the use of a variety of containers from which to pour and measure sand or to mold and shape sand into various forms. Water should be handy if children are to build with sand, however.

Music Areas

A quiet part of the room, away from other activities, can be established as the music center—a space in which to listen to and to make music. Here, children can listen to a song, operating a CD or MP3 player by themselves. They can also play with whatever musical instruments are on hand. Science concepts abound as children begin to understand sound and vibration as well as to enjoy various musical instruments. Environmental CDs may be added to correspond to current science topics such as the ocean.

Computer Stations

Several computers can be set up with age-appropriate programs that

- teach some skills more effectively than traditional and less expensive methods and materials;
- have the potential to help children develop higher-order thought skills like judging, evaluating, analyzing, or synthesizing information (Wright & Shade, 1994);
- do not emphasize war, violence, or discrimination against women or any racial or ethnic group;
- provide for more than one child to work with a program;
- teach a concept or concepts that children are unable to obtain through an active experience or inquiry.

The Interactive Whiteboard (IWB) is a flexible and versatile learning tool for early childhood classrooms. Young children prefer using the IWB as opposed to a computer because they find a keyboard and a mouse difficult to manipulate (Goodison, 2002). Children who would not usually choose to work on a computer will often choose the IWB. One of the major advantages of IWBs as a learning tool is that they are interactive. Children enjoy interacting physically with the board, manipulating text and images.

Quiet Spaces

Children need space to be alone with one or two others. It may be a corner of the room with a few pillows on the floor, a small nook in the library area, or a chair and table somewhere away from the other centers. Every room needs a space—wherever it is, or whatever it consists of—where children can be away from the group, relax, calm themselves, and reflect on their investigations.

THE TEACHER'S ROLE

Without a concerned, interested, and knowledgeable adult, even the best-equipped indoor spaces fall short of offering children meaningful experiences. Based on knowledge of children and of their experiences at home and in the community, it is the adult who

- selects, arranges, and changes the indoor centers, making sure that the spaces remain uncluttered, safe, inviting, and accessible to all children. Teachers provide the raw materials and tools to inspire questions and aid children in their investigations.
- schedules large blocks of time during the morning and afternoon for active experiences.
- provides a background of meaningful experiences with people, places, and things so that children will have ideas—including the imaginary, feelings, and emotions to express through play.

Most of all, however, it is the teacher who interacts with children in ways that clarify, extend, and expand their knowledge and skills:

- Teachers observe and supervise children as they play. Observations can focus on the total group of children or on individuals. The progress children are making, the skills they are gaining, and the things they still need to learn can be noted. When needed, teachers step in—setting limits, clarifying rules, removing broken toys or objects, and supporting children in their attempts to learn new concepts, skills, and attitudes.

- Teachers enter into joint activities with children, working collaboratively with them on a problem or task, such as comparing seeds from four different and familiar plants.

- Teachers extend children's play by entering into the play scene. A teacher of 3-year-olds subtly encourages children who are playing grocery store by saying things like, "Where can I pay for my bananas?" and "This would be a good place to arrange the cereal." These suggestions lead children to organize and extend their play.

- Teachers use language to promote children's learning, naming things in the children's environment and giving information when needed: "That sign says 'STOP.' When we see it, we stop and look both ways before crossing the street."

- Teachers ask a variety of questions that lead children to new learning: "Let's count the acorns; how many do you have?" "How are they alike/different?" "How are you going to mix the colors for your painting?" "Is your suitcase light or heavy?"

- Teachers offer assistance to help a child solve a problem or achieve the next level of functioning: "Here, I'll hold this piece of wood while you attach that part."

- Teachers plan and help children select activities that are appropriate for each individual child's development and background of experiences.

- Teachers set expectations for classroom behaviors that are consistent with children's emerging cognitive and social capabilities (Berk & Winsler, 1995).

- Teachers demonstrate how to do something, supporting children as they try.

- Teachers give specific directions and information when required by students to continue their investigations.

- Teachers seriously enter into conversations about children's work, focusing on the children, their work, and their ideas. Lilian Katz (1993) observed that teachers often seem reluctant to engage children in meaningful conversations and focus more on giving positive feedback rather than talking about content, relationships, or even what the child is doing.

- Teachers carefully structure their interactions with children to move them forward in concept development through understanding of what Vygotsky (1978) termed the *zone of proximal development.*

- Teachers ensure that all children are able to take part in centers of interest.

SUMMARY

Active children need indoor spaces that are specifically designed to foster active experiences with the sciences. Planning indoor environments begins with making certain that the spaces are healthy, safe, beautiful, and accessible to children with special needs.

Indoor spaces are arranged with centers of interest, but science is integrated throughout. Centers organize children's environment, let them see the choices available to them, and give them the means to work and play cooperatively with others.

As Dewey (1938) suggested, the role of the teacher is more complex and more intimate when children are actively engaged in experiential learning. Teachers need a basic understanding of science, which involves both the understanding of science inquiry as a dynamic process and a familiarity with the science concepts that they will help children explore (Moriarty, 2002). They schedule blocks of time for children's indoor experiences and actively teach, guide, focus, and interact with children.

3

Active Children—Active Outdoor Environments

Young children must have a rich outdoor environment full of irresistible stimuli, contexts for play, exploration, and talk, plenty of real experiences and contact with the natural world and with the community.

White, 2008, p. 9

WHY PLAN FOR OUTDOOR ENVIRONMENTS?

Outdoor environments provide rich opportunities for learning. Children need the challenge and freedom inherent in outdoor play. Through rough-and-tumble outdoor play, children have the opportunity to develop feelings of confidence not only in themselves and their bodies but also in others and in their natural environment. Children with physical disabilities find outdoor play of special value. Here they can strengthen large muscles. Depending on their needs, they can walk up and down hills, climb, and exercise small muscles by digging in the sand or playing in water. As they explore the outdoor environment, all children make discoveries about the properties of their world. Children can also carry out investigations and explorations when they are outdoors.

Outdoor play and exploration are an essential part of the science curriculum because young children learn through their senses—hearing, smell, taste, touch, and vision. They also learn science concepts through the motoric manipulation of objects. Active experiences outside the school walls provide the opportunity for sensation, observation, and cooperation, including the following:

1. **Sensory experiences.** Children experience their world through sight, sound, touch, taste, and smell. "Because of the freedom the outdoors offers to move on a large scale, to be active, noisy and messy, and to use all of their senses with their whole body, young children engage in the way they most need to explore, make sense of life and express their feelings and ideas" (White, 2008, p. 7). The outdoors provides many opportunities for children to use their senses. They can smell flowers and plants, listen to the sounds of the wind chimes, feel the soft sand on their feet, or enjoy the taste of fresh fruits or vegetables straight from the garden.

   Teachers should expand water and sand play out-of-doors. As children play with hoses or run through a sprinkler, they explore the properties of water. They can wash doll clothes and hang them in the sun to dry, or hold a car wash and wash all the trikes, wagons, and wheel toys. Adding a trickling water hose and elastic squirt bottles filled with water to the sand area lets children create large structures in the sandbox.

   Digging in the dirt and extracting different sizes of pebbles and rocks is another sensory experience that provides information to the child about the nature of the surfaces of the earth, concepts of heavy and light, rough and smooth, and large and small. Learning is best reinforced when teachers have questions to focus children on their inquiries. For example, when examining rocks, focusing questions might be, "How are they alike/different?" "Which ones are smooth?" and "Which ones are rough?"
2. **Observation.** The natural world is filled with things for children to observe. Clouds, rain, sprouting seeds from a simple class garden, falling leaves, birds, insects, shadows—all are experienced out-of-doors. Teachers should encourage children to observe carefully and formulate questions to be answered through

active inquiry and discovery. Teachers can record the observations of younger children (who cannot write), and older children can record their observations in a science journal or on chart paper. Woyke (2004) suggests that one important goal in any nature-based education is to foster a child's sense of wonder. Wonder happens when a child rolls a log over and finds a colony of tiny creatures. Wonder cannot be taught, but children develop it through discovery and ample opportunity to investigate the outdoors. Worth and Grollman (2003) describe the story of a teacher of 4- and 5-year-old children in an urban public school. Her class was able to study animals outdoors in their natural environment. The essential element of the story is: The teacher needs to get herself and the children ready to be naturalists. The teacher spent a lot of time in the school yard examining what was to be found. Then she made sure that the children had the tools for their investigations—in this case, hand lenses and Popsicle sticks as tools to pick living things up without hurting them. Finally, she instilled in the children a respect for the living things in the outdoor environment.

3. **Opportunities to cooperate with others.** An important part of children's explorations is telling others what they see, what they think, and what further questions are raised by their experiences. "In order to process, make sense of, and learn from their ideas, observations, and experiences, students must talk about them" (Michaels et al., 2007, p. 88). The *Benchmarks for Science Literacy* (AAAS, 1993) further suggest that when they are in kindergarten, children should learn to work in small teams (rather than as isolated individuals) to ask and answer questions about their environment. Yet—consistent with the findings of Piaget (1954)—it is noted that children learn by reaching different conclusions and working out disagreements about what their findings mean.

   Of course, children play cooperatively indoors, but being outside somehow fosters expansive and often long-term cooperative efforts—for example, to construct large buildings and objects. Complex schemes for rearranging equipment, digging gardens, making cities in the sand, and creating areas for the observation of insects and birds develop and bloom out-of-doors. Suggestions for many of these activities are found in the following chapters.

SPACE PLANNING

The out-of-doors is really an extension of the stimulating, well-arranged indoor learning environment. The added richness of natural surroundings and open spaces enhances possibilities for active science learning experiences. Yet science activities are intimately connected and integrated with all areas of the early childhood curriculum. "This integrated approach to science education weaves physical, sensory, and emotional activities into the total learning process" (Harlan & Rivkin, 2004, p. 14). The following outdoor areas provide opportunities to integrate science experiences with other areas of the curriculum.

Science and Nature Discovery Areas

Spaces should be designated for exploration and discovery. Some suggestions include the following:

- An area set aside for bird feeders, birdhouses, and birdbaths. Children can observe and record from a distance the comings and goings of a variety of birds, their nests, and the foods that they eat. If they are lucky, it might be possible to observe eggs hatching. The teacher should encourage children to listen carefully to the sounds in this area of the play yard. Of course, the type of birds observed will vary with locale and weather conditions.

A child learns about a sea creature with her hands.

- An outside sandbox. Although an indoor sand table provides children with many active experiences with science, an outside sandbox has many advantages. Children can climb into it, sit in the sand while playing in it, and have the space to create on a large scale. The sand area should be adjacent to a water source. Props for discovering the properties of sand—such as colanders, sieves, funnels, small containers, cooking utensils, and sand molds—should be added.

- A carefully supervised wading pool or water table. Water play gives children essential skills in pouring, measuring, and comparing. Outside, much more splashing and spilling is acceptable, and water can be carried around the play yard. Many items may be added as required by the type of investigation children are pursuing. For example, water and bubbles illustrate the refraction of light.

- An area where insects and small animals flourish around bushes, plants, and flowers and in the dirt. The life cycle of living things can be directly observed and recorded in photographs, drawings, and graphs. Watching birds, insects, and small animals fosters the concept of the variety of life on earth.

- A garden area for various types of plants, flowers, and vegetables. This space should be carefully placed beyond the limits of children's active play and close to a water source so that very young children can be involved in nurturing the plantings. Some teachers may want to reserve part of the gardening area for impromptu digging in which children learn about the composition of soil, the insects and worms that make their homes there, and the organic matter that decomposes to produce soil. The area should be arranged to receive at least 6 hours of sunlight a day. In locations where this is not possible, teachers should plan gardens that flourish in the shade. As at the Center for Young Children, the process of planning and planting a garden can become an investigation spanning months. The garden area can also include a compost pile or composting bin. When composting, children make a pile of organic waste such as decaying leaves or grass mowings. From time to time, they can then observe as the compost decays and turns into rich soil and dig up shovelfuls to use to help their plants grow.

- An area for exploring sound. This space could include an assortment of carefully arranged wind chimes. Different types of large-scale musical instruments made from pipes and other materials can be made available for children to investigate sound. Other materials can be added as dictated by the type of investigations being conducted by the children. For example, everyday objects such as tins, plastic bowls, or wooden blocks can be used as drums.

The process of planning and planting can become an investigation.

- The outdoor area could also include materials for children to use to investigate the weather. A weather station can include materials such as a rain gauge, windmills, streamers, windsocks, and sundials. The teacher can encourage children to observe the weather conditions and to use simple instruments such as the rain gauge or windmill. Then children could record their observation using different sorts of symbols.

These suggestions offer a planned approach to using the school yard as a science discovery center; they may be supplemented by carefully chosen field trips. The most effective learning experiences probably occur in the natural environment, which is discussed in Chapter 14.

Art Activities

Any art activity can take place outside. Whether it's painting with water on school walls or sheds or painting with color on large brown paper strips hanging on a playground fence, children and teachers can enjoy aesthetic experiences with the knowledge that cleanup will be simple. Further, children learn science concepts by experimenting with the properties of water, observing under what conditions water dries, and mixing paints to the desired consistency or colors to the desired hues.

Children can draw on the hard surfaces with large chunk chalk (a form of rock) dipped in water. Their creations will wash away with the next rain. Or they can use crayons or markers on large papers spread on the yard or tables. Modeling with clay and other materials is fun outside, as is building with boxes, found objects, and other materials. Through their constructions, children learn about the attributes of things and discover which are effective and ineffective in creating satisfactory structures. Measurement and number are involved in creating structures that work, and when things fall apart, children can make inferences about suitable and unsuitable choices.

Math Activities

Math skills are intimately intertwined with those required for grasping science concepts. Outside, it is possible to use numbers to count things and place them in order. Children may estimate quantities of rocks or trees by using their observational skills. The smooth stones that a child gathers, the acorns another collects, the sticks or plastic cups in the sandbox, the number of children waiting to ride a new tricycle—all these give children something concrete to count. Children also classify the stones, insects, seeds, and acorns they find or place them in order from smallest to largest, heaviest to lightest. Shapes such as circles, squares, and triangles can be used to describe many things that can be seen in the play yard, in the natural environment, or in buildings viewed from the outside.

Children learn about math and the physical world through materials, small pieces of equipment, and props that can be moved around and with which they can build and be creative. With movable equipment, children can construct their own play environments using boxes, boards, barrels, tires, and tubes.

Physical Activities

Space and a variety of stationary and movable equipment foster running, jumping, and climbing out-of-doors. Experiences with simple machines are also abundant outdoors. Wheeled toys can introduce the principle of the wheel; a seesaw can introduce the lever. Equipment that promotes social interaction, use of language, and cooperative play and that is rich with potential for children to form concepts of the physical properties of their world includes the following:

- Large wooden crates and boxes, boards with cleats, and large hollow blocks. The size of these materials demands that two or more children work together to use them for building.
- Climbing equipment that comes in several movable sections, such as a trestle unit, a climbing gate, or an A-frame unit, that can be arranged and rearranged to meet children's changing interest or needs. Playhouses often come designed with storage sheds below.
- Cable tables of assorted sizes, tree trunks with sharp branches removed, and sturdy wooden barrels.
- Balance beams, an old log, several logs placed end to end, a board placed on its side, or stepping stones, patio stones, or old tires placed in a series give children a sense of different ways to handle their bodies in space.
- A wide assortment of balls of all sizes and weights.
- Things to push, pull, and ride give children a sense of control over their environment and assist them in learning how things work.
- A wooden ramp and a bucket of balls of different types and sizes assist in investigations of gravity and how various conditions affect it.

Other Considerations

Ideally, outdoor environments should include sections with different surfaces such as grass, concrete, earth, and sand. Pine bark, cedar chips, or pine needles can be used to create soft areas, and other surfaces can be paved for playing organized games, which gives children a sense of using their bodies in space. Play yards should have a balance of sun and shade, active areas for large motor activities, and quiet areas where a child might take a book, listen to music, reflect on the outdoor environment, or play with small manipulatives. Attention should be given to different terrains. A balance of levels can make the play yard visually interesting and provide hills for climbing and rolling down as well as flat surfaces for organized activities. Outdoor pathways wind around the out-of-doors and define various interest areas. They can be created out of cedar chips, stones, or cement. Yet, as the following chapters illustrate, children can learn the techniques of inquiry and experience a sense of wonder in a downtown urban environment with limited resources.

The *Benchmarks for Science Literacy* (AAAS, 1993) highlight the importance of tools to scientific inquiry. Teachers should add tools such as thermometers, magnifiers, and rulers to aid children in their scientific observations. Simple cameras provide an excellent way for young children to document discoveries in the outdoor environment.

General Safety and Health Considerations for Active Outdoor Environments

It is important to ensure that the materials, tools, resources, and equipment provided for young children are safe. The suggestions for safety considerations listed here should be used with a view to enabling rather than preventing children from engaging in active explorations and investigations:

- Fencing at least 4 feet high surrounds the play yard.
- All surfaces are properly maintained and checked for debris such as broken glass and animal waste.
- All plantings within the fence are nonpoisonous.
- Play equipment is inspected daily for missing or broken parts, splinters, and sharp edges.
- Sandboxes have a retractable cover.
- Play area is well drained—water is not allowed to sit.
- Soil is tested for lead and other toxic substances and fresh soil is brought in for digging and gardens when necessary.
- Climbing and sliding equipment is anchored firmly in soft ground cover at least 1 foot deep.
- Equipment is developmentally appropriate (or can be adapted) for the age groups using it.
- Railings enclose high equipment to protect from falls.
- Equipment is spaced for safe movement between pieces.
- Play yards are accessible to children with disabilities, foster their independence, and enhance skill development.

THE TEACHER'S ROLE

Some teachers operate under the mistaken assumption that they are free from any responsibility for children's learning when the class is outside. Yes, there are supervisory and safety considerations, but children are now on their own to let off steam and have a good time. These teachers see outdoor play as a break for them and an opportunity to talk with other adults. "Changing an educational mentality in which children are simply 'let out for recess' and nothing more, thus fostering aggressive unsafe play, is difficult when funds for education are shrinking and other conspicuous needs are emerging. Teachers will make the difference, asserting the essentialness of a safe and challenging outdoors for education and weaving outdoor experiences into their academic curriculum" (Rivkin, 1995, p. 53).

The early childhood science curriculum, perhaps more than other curriculum areas, requires that the teacher take an active role in facilitating children's active sciencing through planning and using teaching strategies that "push" children's thinking beyond its current level. Vygotsky's theory suggests that children can be guided by explanation, demonstration, and appropriate questioning to higher conceptual levels through interaction with more capable and competent adults and peers.

Thus, in the outdoor (as well as the indoor) environment, teachers

- engage children actively. They provide many and varied opportunities for observing, communicating, comparing, organizing, and inferring on the basis of active outdoor experiences.
- serve as models displaying the attitudes and skills they want for the children. "Teachers who think that holding worms or mice or observing spiders is distasteful must work to overcome such aversions or to incorporate the content with different examples, such as caterpillars, gerbils, and ants. Teachers are not expected to know all there is to know about sciencing. What is important is that the teacher be open, enthusiastic, and willing to wonder 'What happens if . . . ?'" (Kilmer & Hofman, 1995, p. 43).
- enter into joint activities with children, working collaboratively with them on a problem or task, such as building a bird feeder or charting the number and type of birds that are using the feeder.
- assess children's attitudes, skills, and knowledge in the area of science using multiple methods. Through observations of children's work and focused discussions with children, teachers plan for outdoor experiences at the next level of thinking.
- employ appropriate questions and comments to focus children on their scientific inquiries.
- document the children's work continuously. As mentioned earlier, a digital camera is excellent for that purpose because it permits duplication of photographs for books or to be given to individual children and their families.

Charts of all types should adorn the walls. For example, "What do we know about fall?" "What would we like to know?" "How many children enjoyed tasting the pumpkin/did not enjoy tasting it?" "Which type of baby food did you like best?" "What is similar/different about the rabbit and the hermit crab?"

SUMMARY

Active outdoor environments for young children enhance the possibilities for learning in all curricular areas; however, natural surroundings and more open spaces are especially important for active sciencing. Well-planned outdoor environments fulfill important benchmarks for learning by providing young children with sensory experiences, opportunities for observation, and opportunities to cooperate with others and to ask and answer questions about their environment. In science and nature discovery areas, places are planned for observation of birds, insects, and small animals, and exploration of sand and water. Because not everything can be planned outdoors, incidental learning plays a large part, such as when a different type of bird visits the bird feeder in the fall.

A variety of stationary and movable equipment fosters children's physical activity and understanding of how their bodies move in space. Safety, aesthetics, and a balance of environmental factors are taken into consideration when planning for outdoor play and learning. A wide assortment of props and tools aids children in their play and scientific observations. Yet no outdoor environment is a learning environment without a teacher who is active in planning and encouraging children's concept development.

4

Building Connections to Home and Community Through Active Experiences

Wonderful resources to enhance science education outside the classroom are flourishing in communities throughout the country. Interactive children's science museums, nature centers, zoos, aquariums, and planetariums offer informal science programs for children and families in pleasurable, noncompetitive settings.

Harlan and Rivkin, 2004, p. 26

According to *Ready, Set, Science!*, "Experiences outside school influence and shape knowledge and skills that children bring with them to the classroom" (Michaels, Shouse, and Schweingruber, 2007, p. 7). John Dewey believed that schools could not function "when separated from the interest of home and community" (Dewey, 1944, p. 86). He depicted the school at the center with the free interplay of influences, materials, and ideas flowing to both the home and the natural environment around the school building and back again to life in school. The work of Piaget (Youniss & Damon, 1992) and Vygotsky (Berk & Winsler, 1995) further confirms the importance of building connections to home and community through active experiences.

According to Piaget, "culture-acquiring" children construct knowledge about their world by interacting with the environment, peers, and adults in school and beyond the school (Youniss & Damon, 1992). Vygotsky (1978) saw children as learning to think and develop concepts by mastering challenging tasks in collaboration with more knowledgeable members of their society. In addition to teachers, our communities have many specialists that may be available as resources for classes and for individual students. Moreover, many communities have access to science centers, museums, national laboratories, and industry that can contribute greatly to the understanding of science and encourage students to further their interests outside of school. In addition, the physical environment in and around the school can be used as a living laboratory for the study of natural phenomena. By using the outdoor environment, teachers allow children to "learn science by actively engaging in the practices of science" (Duschl et al., 2007).

Outstanding Science Trade Book for Students award-winner *Nature in the Neighborhood*, written and illustrated by Gordon Morrison (2004), is a valuable resource for teachers and children in an urban setting. It describes the animals and plants found in the urban environment and how they change with the seasons. Focusing on plants and animals in a single neighborhood in an unnamed North American city, the author encourages children to look closely at common neighborhood scenes such as an alley, a vacant lot, a family garden, and train tracks, which supply habitats for a robin to nest in a fire-alarm box and monarch butterflies to lay eggs on milkweed plants. It is not necessary to visit the rainforest to find resources for science inquiry.

Working with others in their school and with the community, teachers should build these resources into their interactions with children. Thus the immediate environment of the school and community serves as a laboratory or workshop for children to discover the world around them and the people and things that populate that world. Without these active experiences, children lack the raw materials to construct learning in the classroom that has meaning and integrity. Children are cheated when teachers provide pictures of birds, insects, and trees and expect children to form concepts from these instead of observing them in the world outside the school. In addition to observation, in the natural environment, children may engage in the science process skills of comparing, counting, classifying, defining, communicating, hypothesizing, predicting, testing, and experimenting.

In his ecological approach, Bronfenbrenner (1979) placed the developing child in the center of a series of interlocking settings. Home, school, and the neighborhood serve as the immediate basis for child development and learning. Just as active children derive meaning from their experiences in the classroom, that meaning is extended and broadened when teachers recognize the family as a resource for deep and personally meaningful learning experiences.

Mr. Porter, a teacher of 4-year-olds, had attended some summer workshops on the teaching of science and became convinced that it was necessary for him to build connections with the home and wider community through well-planned excursions. Using his experience from the workshops and the help of his colleagues, he formulated the following criteria for experiences with integrity and meaning:

- There is a continuity of experience as one builds upon another. Mr. Porter had school polices that prevented him from taking a field trip each week to a different place, and he realized that the children were having a series of isolated experiences that were soon forgotten. Building experiences instead around one theme or concept area would help children construct knowledge in depth and generalize it to other areas.
- Each experience is worth the child's and teacher's time and effort. A trip to the university laboratory had been long and exhausting. The exhibits had been placed too high for most children to see or touch. Worse, the university student in charge was not used to young children and lectured them as she did secondary students. The children whined, fidgeted, and started poking and fighting with one another. Later, when asked what they remembered about the trip, the children talked only about the drinking fountains.
- Advanced organizers such as T-charts, Venn diagrams, KWL charts, concept maps, and webs should be developed with input from the children. Children need opportunities to discuss, read about, and role-play the excursion in advance. Some teachers find that asking children to make a list of what they already know, what they want to see, and what they think they will see on the trip is helpful. After a walk in the park, for example, children could compare their list with the actual experience.
- Children should have time to reflect upon and follow up on experiences with plans and projects that enhance and expand their learning.
- A planned experience should either be an outgrowth of children's deep interests or meet a specific need for the children in learning subject matter content. One of Mr. Porter's colleagues was teaching her class of 3-year-olds about electricity. They seemed to be having a good time playing with small bulbs, batteries, and wires. Yet when asked what they were learning, they told him, "Magic." At first confused, Mr. Porter soon understood that this abstract concept (electricity) had not been made concrete for the children and that the activities did not have personal meaning for them.
- Flexibility is essential. Although most outside experiences are well planned in advance, teachers can capitalize on incidental learning, such as when children spot a nest of baby birds, find a spider web in the outside playhouse, or find interesting and colorful rocks.

OUT INTO THE SCHOOL

Where can the teacher and children find the active experiences that build connections to home and community and take them outside the classroom walls? Whether in a kindergarten, a Head Start program in a large elementary school, a child-care center, or a small cooperative nursery school, there are many meaningful ways to utilize the immediate environment of the building itself and the grounds that surround it.

Inside the School Building

There are many active experiences for children taking a walk around the school building. Children are intensely interested in machines and tools that help us do work. Children who are in an elementary school setting can meet with the secretary, the nurse, a custodian, an office worker, the people who pick up the garbage, and the cook. The teacher might assist the children in preparing a list of questions to ask by getting them to think about the jobs that people do and what tools they need to perform their tasks. Large and small machinery may be observed and described. Children may be asked to speculate what different kinds of tools are used by those who work in the school, such as hammers, saws, pliers, screwdrivers, nails, scissors, staplers, eggbeaters, thermometers, scales, flashlights, and word processors. These learning experiences can be expanded in the classroom as tools are provided for children in various centers of interest.

Outside in the Natural Environment

Studies of children's knowledge of the plants and animals in their environment reveals that they know much less than older people in their culture group, although Native American children were more knowledgeable than Anglo American children (Rivkin, 1995). The immediate outdoor environment of the school provides children with a rich laboratory for studying nature and their physical world. They should be encouraged to think of all the things they can observe and be prepared by the teacher to observe closely and to gather and record data on flowers, trees, small animals, weather, sounds, and seasons. "The outdoor classroom offers an ideal setting in which students can delve into important environmental issues, and develop an appreciation of their natural surroundings which offer an authentic vehicle for creating an enhanced sense of stewardship and appreciation for the natural world" (Rich, 2010, p. viii).

OUT INTO THE NEIGHBORHOOD AND COMMUNITY

Ms. Smith, a teacher of 3-year-olds in a child-care center, had heard how wonderful the regional park was. Yes, it was a 45-minute trip, but it would be worth it for the children to experience the small farm and petting zoo and talk to the rangers.

- She planned to entertain them on the bus with finger plays and songs, and the assisting parents would enjoy it too. It started well. Everyone was enjoying the ride and began singing "The Wheels on the Bus." That is, until several children became ill. Things went downhill from there. Children cried. The adults, exhausted from trying to comfort the children, became cross. On the return trip, everyone was anxious to see the familiar grounds of the center.
- What went wrong here? First, was the trip meaningful, or was it a waste of valuable time and resources? Ms. Smith thought hard about these questions. She soon understood that the trip was not appropriate for the developmental level of the children in her classroom, although it could have been an excellent experience for older children. Her 3-year-olds could find many experiences with life sciences right in their own play yard.

- In planning for meaningful experiences for children, teachers prepare the children, but they also prepare themselves. The purpose of a trip is to provide children with firsthand experiences based on their interests that they would be unable to have in the classroom, in school, or on the immediate grounds. In fact, the first step for the teacher might be to decide whether the purpose could be accomplished another way. Do the children need to go to the science museum, or would it be possible to bring science resource persons into the school to plan experiments and pose problems? If school visits are not possible, the teacher decides on the goals and then plans for experiences prior to, during, and after the trip.
- Teachers should become familiar with the community and its resources prior to planning any trip. Unforeseen difficulties can be avoided if teachers preview the site and talk with the people at the places they wish to visit, such as the nature center or the children's museum. Some sites such as museums and libraries have prepared tours and materials for children; however, these may be too long or complex for very young children. Teachers should shorten and modify an experience when necessary and create their own materials that will be age-appropriate.
- Teachers should also consider the integrative power of a field trip (Seefeldt, 1997). How will the science trip facilitate growth in the language arts, mathematics, and the use of books, writing, the arts, and social skills? As teachers prepare the children, they will emphasize active experiences in all of these areas. In addition, they will choose children's literature that integrates science throughout the early childhood curriculum. Each active experience will involve learning new vocabulary words and investigating the site through informational books and fiction. Children should draw pictures, dictate stories, and sing songs about what they have done. They will re-create and reinvent their learning through dramatic play inside and outside the classroom. Social skills develop as children experience new people and places and acquire behaviors to fit the situation.

BASIC GUIDELINES FOR MEANINGFUL FIELD EXPERIENCES

1. Keep the experience simple for very young children, and increase the complexity as it is developmentally appropriate. For example, very young children will profit from a short walk to collect leaves in the fall. With the teacher's help, they may classify the leaves first by color and then by shape. Older children will advance their concept of the relationship between food and health by visiting the food store to select items that correspond to the sections of MyPyramid for Kids (USDA, 2005).
2. Consider the mode of transportation. Walking is best for most children, yet some field trips necessitate bus or public transportation. In fact, when learning about gears and wheels, a trip on the bus is an authentic and active experience.
3. If the classroom is inclusive, consider all aspects of the field experience. Pathways and sites must be barrier-free and experiences must be open-ended so that all children profit from the trip. Small-group excursions may provide a better opportunity for all learners to profit.
4. Introduce the field experience through discussions, pictures, reading about the things to be viewed, and art experiences. Ordinarily, field experiences will be part of a larger scientific investigation, such as gardens. For example, if children are to visit botanical gardens, they may observe, compare, and describe the many plants they find in the play yard; plant a seed and observe the effects of water and light on its growth; cut pictures from magazines of beautiful or

exotic plants and make a collage; read about plants that grow in different types of soil; and even vote on their favorite plant. Small-group work in areas of particular interest should be encouraged.

5. Organize play around the places to be visited. The creative dramatic center and the outside environment can be adapted to fit science field experiences. Often additional clothes or props will be needed. Young children wear lab coats and carry simple equipment for collecting specimens for science classes at the Lawrence Hall of Science in Berkeley, California.
6. Prepare the children to observe closely and gather data during the field trip. Their observations will be used as the basis for many activities in the days and weeks to follow. For example, children have probably been on walks many times with their parents. Yet they have not paid close attention to the smells, shapes, variety, and colors of the trees and others plants that line their way. Questions originating from older children may be compiled on sheets to remind them of the experiences they wish to have.
7. Give children plenty of opportunities to reflect on their experiences. Allot time and materials for follow-up plans and projects. Isolated experiences are easily forgotten. Learning from field trips is part of an integrated curriculum.
8. Document children's experiences through photographs, exhibits of children's work, and charts reiterating the process of the investigation.
9. Welcome parents during any phase of the planning, implementation, or follow-up. Opportunities for parent participation should accommodate parents' schedules. Parents need not come to the school or field experience. There are many ways in which they can enhance their children's experiences in the home if they are informed of the teachers' plans and activities.

Children prepare for a study of fall by recording what they know.

Some Safety Tips for Field Experiences

- Obtain parental permission for children to participate in the excursion.
- Check the environment ahead, both inside and out, for any hazards.
- Be sure that all teachers and staff members are trained in first aid and CPR.
- Include at least one person with such training on the trip.
- Take a first-aid kit on the excursion.
- Take an up-to-date list of emergency phone numbers for each child.
- Check medical forms for children's allergies (including reactions to wasp and bee stings and foods).
- Always walk on the left, facing traffic.
- Be sure that children understand that small wild animals are for observation and not handling.
- Educate children about poisonous plants and drinking water from streams in the outdoor environment.
- When utilizing transportation, make sure the children know, have practiced, and will follow the rules.
- Consider adult/child ratios. Include no more than three or four children on a field trip for each adult present and fewer if the trip requires complex arrangements for transportation.
- Remember that small-group excursions may be best for all learners.
- Be sure that the field trip site meets guidelines for children who are developmentally different.

BUILDING CONNECTIONS WITH THE NEIGHBORHOOD AND COMMUNITY

According to Dewey (1938), when learners reflect upon and analyze real experience, they are able to form new concepts. They can then test the new concepts and use them in future situations and experiences. The neighborhood and local community offers an ideal opportunity for this type of learning. Young children can engage in real-life problem solving while providing service to their communities (Lake & Jones, in press). For example, adopting a stream or a playground gives children a sense of responsibility and participation in society through real-life experiences. In our throwaway culture, young children can learn environmental values. An excellent resource for teachers who are concerned about the environment is David Burnie's *Endangered Planet* (2004). Children will be drawn to the full-color photographs throughout the book. Additional ways to introduce children to science through community responsibility and caring include visits to a veterinary clinic and carefully facilitated intergenerational contacts.

- Mr. Vasquez noticed that the 5-year-olds he taught were treating classroom animals carelessly. He decided to plan several trips to the local veterinarian to foster empathy and understanding in children. He believed that in the process of learning how to care properly for a small animal, they would expand their knowledge and develop respect and appreciation for all living things. He noticed improvement at the outset, and by the end of the visits, children were volunteering to care for the classroom animals and setting careful rules.

- Through carefully planned intergenerational programs, children and elders interact with one another, reestablishing the relationships of caring and continuity of life that have been broken by physical and social distance. Teachers should plan programs and contacts carefully because children tend to hold negative stereotypes about the elderly and in some instances fear contact with them. With this is mind, it would be unwise to take children to nursing homes, where elders have symptoms of senility and infirmity.
- Seefeldt (1997) suggests that when children are exposed to elders who are healthy, happy, active, and fulfilled, they can share the love of an older person: "Intergenerational programs in the school can provide a way for children and elders to enjoy one another's company, to learn from one another, to share feelings of affection, and to provide children with a concrete example of life's continuity" (p. 195).
- Every community has natural resources, people resources, and material resources. Teachers should make an effort to become acquainted with these resources to extend and expand the science offerings possible in the classroom.

Natural Resources

Parks and nature centers offer extra glimpses into the world of woods, trees, plants, ponds, lakes, beaches, animals, birds, and the like. The National Park Service runs a number of centers that have hands-on science activities for children, trails to follow with guides written for children, and knowledgeable rangers to pose and answer children's questions. The ranger or guide may have a snake on hand or an active animal shelter to show. Local communities also sponsor park and recreational services for children.

Visitors

Visitors who enhance the science curriculum include parents and other specialists in transportation, health care delivery, computer technologies, music, art, cooking, mechanics, and many other fields that have scientific aspects. They can be effective resource people if the teacher prepares the children carefully for their visits and encourages them to provide active demonstrations or hands-on opportunities.

Zoos

Zoos are controversial in the science and environmental community because they remove animals from their natural surroundings and—until recently in this country—relied on fences and other restraints to separate the animals from their visitors. More recently, zoos have become more aware of concerns for animals and employ more natural barriers, fewer cages, and appropriate habitats for the animals. There are many good zoos with highly trained docents and guides to work with children. At the National Zoo in Washington, D.C., the volunteers who work with children must take a rigorous course of study in both animal life and how to teach life science effectively to children. Excellent written materials enhance a child's visit with questions and comments to promote scientific thinking and invitations to make drawings and write stories about the animals. The zoo staff hangs the children's pictures and stories where they can be viewed by other young visitors.

Science Centers and Museums

Most science centers and museums have planned activities for children including experiments and opportunities to handle small creatures. At the Lawrence Hall of Science, young children are invited to don a lab coat and go down to the pond to collect tadpoles.

The specimens are arranged with the children's names and are observed over time. At the Smithsonian Institution in Washington, D.C., "Don't Touch" signs have been replaced by "Please Handle with Care" in the discovery areas of the museum. Trained persons supervise the handling of fragile shells and nonpoisonous insects.

Other community resources that teachers may want to consider are

- colleges and universities;
- radio or television stations;
- transportation services such as airports and train stations;
- professional services such as doctors' offices and hospitals;
- commercial services such as bakeries, pharmacies, factories, and farms;
- observatories;
- aquariums.

The Home–School Connection 

The family has long been considered the child's first and foremost teacher and possibly the child's primary community for learning (Barbour, Barbour, & Scully, 2010). Teachers can utilize the family as a resource for children's learning both in the school and in the home. There is an increased research base on the benefits to the child of family involvement, even if the extent of that involvement is small (Epstein, 1991; Grolnick & Slowiaczek, 1994; Mine et al., 2010).

- Teachers and parents may face some challenges in working together. According to Barbour et al. (2010), early childhood educators increasingly service families

Parents support the teaching of science by the interactions they have with children at home.

characterized by single-parent households, cultural diversity and ethnic minority status, dual-worker or dual-career lifestyles, economic pressures, and geographic mobility. The new demographics of family structure call into question the viability of existing approaches to relations between families and early childhood programs. Yet frequent contacts cement a genuine respect and tolerance for different family types. Involving parents as active partners in the classroom provides both parent and teacher with firsthand information about the expectations of the home and the school.

- Project 2061 of the American Association for the Advancement of Science has created questions that parents can ask their local schools. These questions were designed to encourage communication among parents, teachers, school administrators, and the community as a whole. They can also be found on the American Association for the Advancement of Science website (http://www.project2061.org/publications/articles/questions/10QText.htm). Similar questions can be formulated for parents of younger children:

1. Is science literacy for all high school graduates a major goal of the K–12 program?
2. What provisions are made in the curriculum for students of different interests, talents, and ambitions to succeed in science?
3. What is the proportion of females and minorities enrolled in advanced science classes?
4. Do teachers at different grade levels work together to clarify what ideas will be learned and when they will be learned?
5. Are students learning connected concepts rather than simply memorizing isolated facts, formulas, and technical terms?
6. Is the learning active and student-centered?
7. Do teachers welcome curiosity, reward creativity, and encourage healthy questioning?
8. Are teachers given encouragement, time, and resources to update their own skills and knowledge?
9. Do teachers look for and deal with students' misconceptions about how the world works?
10. Do teachers and school administrators use national or state standards as guidelines for improving student learning?

Classrooms work best and children learn more when parents are involved. Teachers may employ the following multiple approaches:

- Conferences in which teachers provide parents with samples of children's work and invite parents to share their observations about their children's learning and their suggestions for classroom and community experience based on their children's interests.
- Use of informal contacts. Busy parents enjoy a brief chat before school, telephone calls, informal notes, and bulletin boards that inform them of plans and programs and invite them to participate in a variety of ways.
- Somewhat more formal contacts through the provision of a Parents' Corner or Family Room where parents may interact around learning materials and other activities of interest to the family. These often include a lending library with toys and picture and reference books and children's science magazines such as *Wild Animal Baby,* which is published by the National Wildlife Federation.
- Regular newsletters explaining the goals for the week, why certain activities were planned, and how parents can support the lessons at home. Other items to include

in newsletters are special events at school, science activities that children have enjoyed, special television programs on scientific topics that parents and children might watch together, and special events for children and families occurring in the community.

- An open-door policy for parent observation and participation. If parents are unable to work regularly as paid or unpaid volunteers, they may make or send materials for special projects in the classroom, help on field trips, or come to school when their schedules permit.
- Selection of active science experiences for children that can be documented. Children will have products such as drawings, charts, or stories to communicate to busy parents what they are doing in school. Digital cameras allow teachers to make a copy of an important science discovery for each child's family.

Parents support the teaching of science by the interactions they have with their children at home. Brewer (1998) suggests that teachers "help parents understand that children learn science when they wash a greasy dish, water a garden plot, ride their trike down a sloped surface, and so on. Some parents think science experiences have to be formal and difficult for learning to occur" (p. 344). Teachers should emphasize safety issues with parents before they undertake even the simplest activities. Additionally, by learning about the diverse cultures of their students' families, teachers can avoid science activities that will be offensive, such as experimenting with certain foods or studying animals that have symbolic meaning in a culture.

Parents can encourage scientific thinking in their children by asking open-ended questions and taking time to encourage the answers. They can help their children to observe ("What shapes do you see on that tree?"), classify ("Let's put away your toys by color"), predict ("How long will it take that squirrel to get up the tree?"), and quantify ("How tall is that building?"). It is possible to practice the skills of science everywhere (Jones et al., 2008). In addition, it is never too early for parents to encourage children to care for the environment.

Teachers should plan family science events at school and send suggestions home in the class newsletter for supporting science learning in the home. Successful school science events involve the whole family in collecting data, undertaking investigations, and solving puzzles and problems. These are active experiences for the whole family. The standard school science fair usually involves complex materials and equipment and seems more like a competition than a joint exploration that is both fun and intellectually stimulating. The following interactions with parents should introduce them to and involve them in children's science experiences:

1. In the class newsletter, have a weekly suggestion box designed for supplementing the science curriculum at home. For example, highlight free activities occurring in the community, exhibits to be visited, activities based on kitchen science, and backyard activities.

2. Have parents and children work together to create mini-museums at home that can be shared with the class when completed. Send parents a note explaining the goal of the activity, how it is related to work at school, and simple directions for making the mini-museum. An example might be a collection of shells from a visit to the beach.

3. Teachers may want to put together a "science backpack" that children take home on a rotating basis (Brewer, 1998). The backpack should contain a note explaining the purpose of the activity, an information book related to the activity, and all the materials necessary for completing the activity. The teacher should make sure that all of the materials contained in the backpack are safe, easy, and fun for parents to use and are translated into their native language.

Additionally, even if family members cannot volunteer in the classroom on a regular basis, they can share their talents, occupations, hobbies, customs, and traditions with the class and school community as they are able.

SUMMARY

Building and maintaining connections with home and community provides benefits to all. Young children need active experiences consistent with their participation in science education. These experiences require careful planning by teachers, who are rewarded by observing children's authentic learning, as one experience builds upon another into an integrated whole. As the family is recognized as a valuable resource for learning, parents and teachers feel mutually supported and learn to understand and value each other as contributors to the child's understanding of science. All of this is consistent with a strong emphasis on sciencing in the curriculum for young children.

5

Experiences and Science Content

Students learn science by actively engaging in the practices of science. A classroom environment that provides opportunities for students to participate in scientific practices includes scientific tasks embedded in social interaction using the discourse of science and work with scientific representations and tools. Each of these aspects requires support for student learning of scientific practices.

Duschl et al., 2007, p. 342

Firsthand experiences in their classrooms, homes, and communities form the foundation for children's learning. These are like the first rungs on the learning ladder. But unless a firsthand experience leads children to an ever-expanding world of previously unfamiliar facts, information, and knowledge, children's learning will be limited. Thus, the next step on the learning ladder is the expansion and extension of the knowledge that children have gained through their firsthand experiences into a fuller, richer, thicker, and more organized form (Dewey, 1938). This form will gradually approximate the experts' understanding of a given subject matter or content area.

KNOWLEDGE OF CHILDREN

Knowledge of some of the universals of children's growth, development, and learning is necessary for the teacher to capitalize on and plan experiences that are both developmentally appropriate and challenge the limits of children's understanding. There is a wide variation, however, in children's growth, development, and learning. Some is due to the variation in the sociocultural context of children's lives and some to the variation in patterns of normal growth and development. However, research evidence increasingly shows that children are much more capable than was once thought (Duschl et al., 2007).

Even though each preschool child is unique, we know from research and theory that children who are approximately the same age view the world around them from similar perspectives and can profit from science content that is geared to their general level of development. Children come to school with a great capacity for learning and can engage in very sophisticated scientific thinking in the early grades (Duschl et al., 2007). Duschl et al. (2007) claim that "by the end of preschool, children can reason in ways that provide helpful starting points for developing scientific reasoning. However, their reasoning abilities are constrained by their conceptual knowledge, the nature of the task, and their awareness of their own thinking" (p. 53). For this reason, progress in science learning should be developmental. Most of the basic skills and concepts can be introduced to preschoolers and expanded upon through kindergarten and the early primary grades. "Children who have many interesting, direct experiences over time with science concepts will gradually understand the broader principles as they develop the cognitive skills to make more abstract generalizations" (Kilmer & Hofman, 1995, p. 47).

KNOWLEDGE OF THE SUBJECT MATTER

There are many resources that teachers can use to gain a better understanding of children's general growth and development. Authorities in the field of early childhood education—such as Bredekamp and Copple, who edited *Developmentally Appropriate*

Practice in Early Childhood Programs (2006)—describe universal patterns of children's growth, development, and learning. This universal information about children's growth and development informs teachers about the potential and the vulnerabilities of young children; it also tells us that in general, the younger the child, the more wedded he or she is to learning through firsthand interactions with the environment and others. Children who are ages 3, 4, and 5—still in Piaget's preoperational period of cognitive development and busy actively constructing their own knowledge—must rely on firsthand experiences in order to learn. As they grow and mature, however, they become increasingly more reliant on interpreting and extending their experiential learning through symbols, through pictures, and through spoken and written language. They also learn by interacting with their peers and with adults.

Knowledge of the Subject Matter: Science

Knowledge of children alone is not enough if teachers are to expand and extend children's learning. If teachers are to take children further up the learning ladder, then they must also have an understanding of the subject matter they want children to learn.

Just as authorities in the field of early childhood have identified universal principles of children's growth, development, and learning, authorities in the field of science have also identified general facts, information, and knowledge key to science for young children (AAAS, 1993; NRC, 1996). Knowledge of these key concepts, themes, and standards that serve to organize science education guides teachers in the selection of firsthand experiences will serve as a base for children's learning, as well as lead them in expanding and extending these firsthand experiences into more formal, conventional knowledge.

Recently, science educators have redefined what it means to be proficient in science. In the National Research Council publication *Taking Science to Schools* (Duschl et al., 2007), four strands of science proficiency are identified:

1. Know, use, and interpret scientific explanations of the natural world
2. Generate and evaluate scientific evidence and explanations
3. Understand the nature and development of scientific knowledge
4. Participate productively in scientific practices and discourse

The four strands are interwoven and are based on the premise that understanding science is multifaceted.

Many teachers report that they feel less prepared to teach science than any other subject matter area (Wenner, 1993). It appears that such feelings result from teachers' misconceptions that science for children requires a command of difficult concepts and facts, expensive and complicated materials, and an emphasis on student memorization and acquisition of facts. This view of science has changed. During the last two decades, a number of researchers have focused on how children learn. Early childhood educators now believe that children are constructive learners, constantly creating and constructing their knowledge about the world based on their own questions (Barclay & Traser, 1999). From this perspective, children are natural scientists. According to Worth and Grollman (2003), teachers need a willingness to roll up their sleeves, become inquirers, and experience the materials they will be using with children. They further need to learn about science through good books and other media resources. These are suggested in Chapters 6 to 14.

Kilmer and Hofman (1995) believe that "the contribution of early childhood education toward scientific literacy is to lay a solid foundation for the continuing development of an interest in and understanding of science and technology by ensuring that every child—regardless of gender, racial or cultural background or disabilities—actively participates in science experiences and views himself as successful in this endeavor" (p. 44). They use the term "sciencing," which refers to the child's active participation

A mural of a garden provides the final step in the investigation of gardens.

in learning about science and points out the emphasis on process. Sciencing is a "hands-on, brains-on" endeavor. They identify the following three goals for sciencing with young children:

- to develop each child's innate curiosity about the world;
- to broaden each child's procedural and thinking skills for investigating the world, solving problems, and making decisions;
- to increase each child's knowledge of the natural world.

An Emphasis on Inquiry

In general terms, "science" is defined as the process of manipulating, observing, thinking, and reflecting on actions and events. The role of experience is primary. The *National Science Education Standards* view science as inquiry and state that "as students focus on the processes of doing investigations, they develop the ability to ask scientific questions, investigate aspects of the world around them, and use their observations to construct reasonable explanations for the questions posed" (NRC, 1996, p. 121). Worth and Grollman (2003) suggest that even when children learn science inquiry skills, subject matter is important: "Choosing meaningful science content for children's inquiry is also important so that children can build a foundation for later deeper understanding of basic science concepts" (p. 156).

The standards emphasize that the teaching of science as inquiry provides teachers with the opportunity to develop student abilities and enrich students' understanding of science while adhering to developmental appropriateness. Young children are able to begin the process of full inquiry, which involves asking a question, completing an investigation, answering the question, forming generalizations, and presenting the results to others. To present the results to others, it is necessary to plan ways that children can organize their findings (Seefeldt & Wasik, 2002). For example, categorize and classify things that children have observed by size, shape, color, texture, or what the object is made of. Children can also classify by type of plant, animal, or insect life and where these live. Then, with the teachers' help, they can create all types of charts and exhibits for display in the classroom. At the Center for Young Children at the University of Maryland, a mural of a garden provides the final step in the investigation of gardens. The following chapters have many examples of documentation.

ORGANIZING CHILDREN'S EXPERIENCES

Organizing children's experiences around some general themes identified by the *Benchmarks for Science Literacy* and the *National Science Education Standards* makes the task of the teacher seem less overwhelming. The following themes, expressed by both documents, serve to organize this book:

1. **Life science and the living environment.** Science programs for young children should provide direct experience with living things, their life cycles, and their habitats. Young children develop concepts of living and nonliving things, the variety of living things on earth, the process of categorizing living things, the behavior and needs of living things (including their environments), and respect for living things.

   The emergent themes of the life sciences are reflected in chapters 7 and 8 of this book. Concepts involved in exploring the natural world are embedded in these experiences.
2. **Earth science and the physical setting.** Young children are naturally interested in everything they see around them—soil, rocks, streams, rain, sand, and shells. Science should include experiences that provide for the study of the properties of earth materials and the discovery of patterns and changes in water, rocks, and minerals. They are intensely interested in the outdoor environment, naturally use it as a laboratory for learning, and enjoy drawing or recording and charting what they see and think, both individually and in small groups. Some experts believe that some concrete elements of space science should be introduced to young children in the context of the physical setting. These concepts are focused on children's explorations of their immediate environment and not on the formal study of the solar system and space.

   The themes that emerge from the study of earth science are reflected in chapters 10 and 11.
3. **Science in personal and social perspectives and the human organism.** Although young children might struggle to understand the abstractions inherent in this category, central ideas related to learning through the senses and mental and physical health provide the foundations for students' eventual understandings and individual and collective actions as citizens. Concepts and attitudes about personal and collective health and nutrition are essential for well-being. An understanding of how one brings in information through the senses expands children's observational skills and permits the processes of investigating the world, solving problems, and reaching conclusions to occur.

   Health, nutrition, and learning through sensory experiences are reflected in chapters 12 and 13.
4. **Physical science, the physical setting and the designed world.** A science program for young children should include many opportunities for children to exercise their natural curiosity in observing and manipulating common objects and materials in their environment. As children explore the properties of objects and materials (size, weight, shape, and color), they can measure these properties using first simple and then more conventional tools. Beginning concepts develop as young children act on objects to produce a desired effect by putting

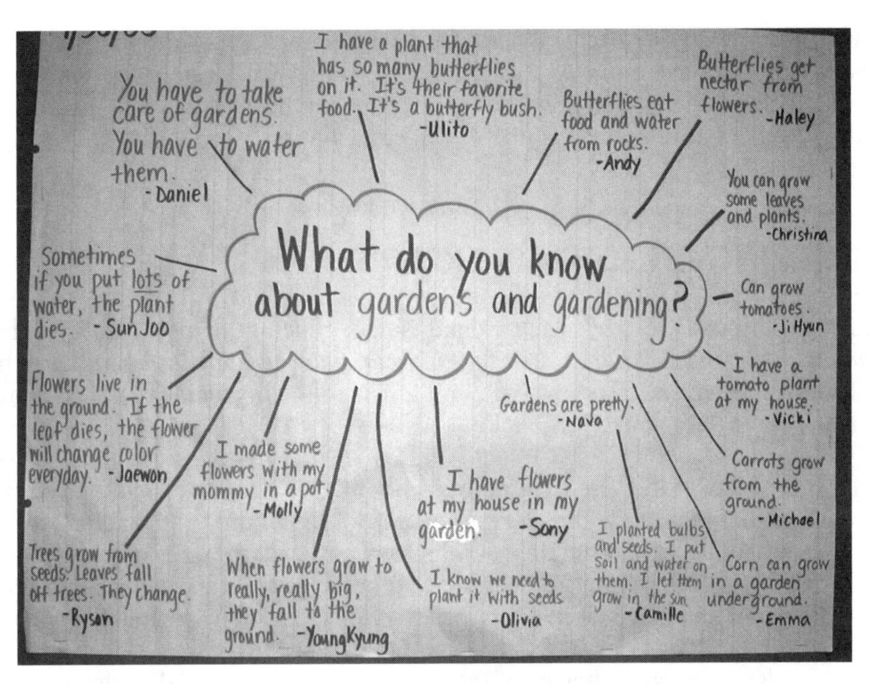

Children use the outdoor environment as a laboratory for learning.

objects together to form constructions of various kinds and draw conclusions about how the desired effect was produced.

Themes that emerge from the physical sciences are reflected in chapters 6 and 9.

BRINGING KNOWLEDGE OF CHILDREN AND CONTENT TOGETHER

To bring children and science concepts together, teachers first need to find out all they can about the science content that they are planning to teach young children. Teachers might follow some of these suggestions in preparation:

- Read informational books for adults and children on the subject they want to teach young children. (See chapters 6–14 and the References section for up-to-date suggestions.)

- Determine the underlying concepts that children can learn. Key concepts can be identified through inquiry into authentic questions generated from student experiences and use of major themes generated by the *National Science Teaching Standards*, the *Benchmarks for Science Literacy*, and knowledge of child growth and development. When a teacher is unsure of a concept and how to make it accessible to young children, he or she might read books on the topic written for students in the primary grades and adapt key concepts for younger children.

- Ask authorities in the field to share their information and expertise. Professors at nearby colleges and universities are often anxious to share their resources with classroom teachers.

- Work collaboratively with other teachers through associations and workshops. Associations such as the NAEYC, the Association for Childhood Education International (ACEI), and the NSTA hold national and local meetings, publish journals and resource catalogues, and have local chapters. There are numerous and topic-specific sessions at conferences on teaching science to young children as well as many journal articles geared toward providing experiences in the life sciences, physical sciences, and earth and space sciences.

- Visit museums, watch videos or television, and search the Web. In recent years, numerous stable websites have emerged for teachers, parents, and children. National associations and science groups are anxious to share their resources with others. The websites are tied to the science content standards and provide ideas for investigations and children's literature to accompany them. See Chapters 6 to 14 for links to the relevant websites.

- Develop a repertoire of questions and comments to promote scientific thinking in children, such as, "Does ___________ look the same today as it did last week?"

- Provide the raw materials to spark children's interests in undertaking a science investigation.

Then teachers need to find out what children already know and understand of the concepts key to science themes. To do so, teachers can do the following:

- Interview children. For example, to find out what they know about water, ask them to
  - tell everything they know about water. Put water in the center of a web and draw lines to children's statements about water. Add the child's name next to the statement.
  - make a KWL chart listing what children know and would like to know (and finally, what they have learned after the investigation).
  - draw a picture about water.
  - construct another web with "water words" or water pictures.
  - explain some of the reasons why water is important to life on earth.
- Take a walk around a child's neighborhood to understand what the child has already experienced and how these experiences can be expanded into conventional knowledge.
- Talk to a child's family, asking about family experiences, what the child is interested in and likes doing, and what the child is not familiar with or needs to become familiar with.
- Observe children as they work and play, noting what themes are involved in their play and how they solve problems, use language, and interact with others and their world.

Finally, the teacher will probably not be able to answer all of children's questions. A perfectly good strategy that supports inquiry skills is to say, "Let's find out together."

EXPANDING AND EXTENDING FIRSTHAND EXPERIENCES

Children's experiences with their world enable them to develop spontaneous, everyday concepts. This everyday, personal knowledge does not, however, automatically lead to a deeper understanding or more conventional ways of knowing. Rather, these concepts act like Velcro, hooking onto whatever new information, facts, and experiences children discover or are given. The richer the new information, the greater the possibility for children to see the relationship of one fact to others and to form generalizations.

Vygotsky (1986) pointed out that at different developmental stages, children learn different things as they independently act on and interpret their environment, but also that other people interact with children, affecting the course of their development and learning. He thought children operated at two levels of thought. One was the stage at which they could solve problems and think without the guidance of an adult or a more skilled peer. The second level was the stage at which the child could perform a task with adult help or guidance. He called this the *potential developmental level*. The distance between the two levels was termed the *zone of proximal development*.

It follows that according to Vygotsky's theory, teachers should first try to understand children's existing ideas and science content. Then teachers can extend and expand children's knowledge by engaging them in activities such as the following:

- Providing children with all kinds of books—poetry, literature, single-concept picture, and reference books—that pertain to concepts children are studying. Books displaying beautiful photographs of insects and small animals come close to providing a child with the real experience. Children's literature is a valuable science tool. It generates interest and motivation, provides context, invites communication, and connects science with the rest of the child's world (Barclay, Benelli, & Schoon, 1999). Books may be openly displayed on a shelf or table, inviting children to extend and expand their ideas. Some of the books children can use independently; others can be read to the entire group or to an individual child or two.
- Looking at pictures and other print media with children. Videos, photographs, movies, slides, and computer simulations of things in—or not in—their environment can be examined and discussed to extend and expand children's knowledge.
- Showing children how they can do something. Teachers, working collaboratively with children, can demonstrate how to plant a seed for optimal growth, observe a rock with a magnifier, or record the progress of a bird's nest.
- Telling children a fact or piece of information that will enable them to make sense of their world. Giving children the proper words for scientific phenomena is something that teachers do to enhance learning. There is no way for children to construct these for themselves.
- Questioning children. Ask children what a thing is, why it is this way, and how it got this way to spur their thinking in a new or different way. Questions and comments are key to promoting scientific thinking. When children encounter new materials and phenomena, teachers may need to intervene to focus and challenge students or the exploration might not lead to learning. Premature intervention, however, may deprive students of the opportunity to confront problems and find solutions by themselves (NRC, 1996).
- Asking children to observe and listen to authorities show or tell about their field: a docent at the Smithsonian Institution can explain the features of rocks and minerals; a park ranger can reveal some of the mysteries of nature during a walk on the nature trail.
- Having children use the computer to learn a new skill or fact, find information, or communicate with others. When utilizing computer software, teachers should apply the criteria of age appropriateness, child control, clear instructions, and expanding complexity.

- Adding another experience. Based on an understanding of children's ideas and concepts in the area of science, add another real-life experience that will expand and extend this knowledge.
- Providing multiple opportunities for children to learn from one another. Children should be able to revisit their existing ideas of the subject matter by freely sharing their view of the world with others and arguing their point of view. Only through interactions with others can children critically consider their existing ideas and revise these to form more complex and conventional concepts of their world.
- Modeling the skills of scientific inquiry. "Teachers who exhibit enthusiasm and interest and who speak to the power and beauty of scientific understanding instill in their students some of those same attitudes toward science. Teachers whose actions demonstrate respect for differing ideas, attitudes, and values support a disposition fundamental to science and to science classrooms that also is important in many everyday situations" (NRC, 1996, p. 37).

By respecting the children, how they learn, and the subject matter of the sciences, teachers extend and expand children's existing knowledge. Teachers teach. They also provide time, space, and materials to promote inquiry. Bredekamp and Rosegrant (1995) ask teachers "not to water down the learning experience even for the youngest child" (p. 22), but rather to build on children's existing knowledge and experience, continuously assessing and supporting learning.

SUMMARY

Firsthand experiences enable children to construct everyday, spontaneous concepts. These concepts are like the first rung on a ladder of learning. The role of the teacher is to extend and expand these concepts into fuller, richer, more conventional knowledge. Teachers do this by developing an understanding of children, how they learn, and a knowledge of science content and then bringing the two together.

Authorities in the field of science education have identified themes or concepts key to the field. These are used by teachers to guide them as they extend and expand children's everyday concepts.

Concepts can be expanded and extended in a number of ways. Books, print media, field experiences, the computer, and other technologies can be made available. Teachers can demonstrate skills, plan investigations, apply concepts, and provide children with the information that will enable them to reach a fuller understanding of the natural world and the scientific principles needed to understand it.

PART TWO

Guides to Active Experiences

6

I Am a Scientist

FOR THE TEACHER

◇ What You'll Need to Know

Digging in the sand, pouring water from one container to another, watching a spider climb the classroom wall—children use all the skills of a scientist. Children *observe* their physical and biological world: "Oh, oh, there's a spider," says Roberta. "I can count its legs—one, two, four," she continues. They *ask questions:* "Where does the spider live?" "Where will it go?" "Why does it have eight legs?"

With the teacher, they *plan* and *conduct investigations:* "Let's get a book about insects," suggests Danny. "Then we can find out where spiders live." Later, after looking through the book, Danny says, "There are no spiders here. What's wrong here?" And so begins a long investigation into spiders as children collect data, listen to experts, research books, observe insects and spiders, and finally *reach conclusions:* "Spiders are not like insects. Insects have six legs; spiders have eight."

◇ Key Concepts

Recognizing that children are scientists, teachers plan to foster children's abilities to:

- observe;
- uestion;
- hypothesize;
- plan and conduct investigations;
- organize their thinking;
- reach conclusions;
- communicate their findings.

◇ Goals and Objectives

The children will develop:

- observation skills;
- the ability to classify;
- the ability to estimate and predict;
- the ability to ask salient questions;
- skills in planning and conducting investigations;
- ways to document their findings;
- the ability to reflect on their experiences and reach conclusions.

◆ Standards Alignment

Key Concepts and Goals and Objectives are based on the *National Science Education Standards* content standards, grades K–4. Standard A: Science as Inquiry (abilities necessary to do scientific inquiry, understanding about scientific inquiry).

Benchmarks for Science Literacy (The Nature of Science)

◇ What You'll Need

Consult books that have resources on the process skills of being a scientist.

1. *Start Young: Early Childhood Science Activities* (2006) by S. McNair (NSTA Press). This book offers a wealth of age-appropriate ideas for teaching science to young children. The chapter with background on the latest thinking about effective ways to introduce science in early childhood is particularly useful.
2. *Benjamin Banneker: Pioneering Scientist* (2003) by G. Wadsworth (Lerner). This biography of a black scientist highlights his almanacs and how he recorded his data.
3. *Learning to Be a Scientist* (2000) by J. Norris (Evan-Moor Educational Publishers). This book is designed for 5- and 6-year-olds, so it is useful for both the teacher and the children. It covers concepts that address the *National Science Education Standards* on inquiry: scientists make observations, scientists sort items and name categories, scientists make comparisons, scientists make measurements, scientists use tools and equipment, scientists record information and explain it to others, scientists conduct investigations.
4. *Worms, Shadows and Whirlpools: Science in the Early Childhood Classroom* (2003) by K. Worth and S. Grollman (Heinemann). This book looks at early childhood science explorations using the words of teachers and students in preschool, day care, Head Start, and kindergarten.
5. *Teaching Science as Inquiry* (10th ed.) (2004) by A. A. Carin, J. E. Bass, and T. L. Contant (Merrill/Prentice Hall). This huge volume is on the philosophy of teaching science as inquiry. The second section has many ideas for teachers based on the *National Science Education Standards.*
6. *A Head Start on Science: Encouraging a Sense of Wonder* (2007) by W. C. Ritz (NSTA Press). This book includes a collection of 89 science activities for children ages 3 to 6 years.
7. *Science in the Early Years: Building Firm Foundations from Birth to Five* (2010) by P. Burton and L. Thornton (Sage Publications). This book provides an overview of the science that should be taught to children from birth to age 5. Especially useful are the examples of science activities and the science concepts, skills, attitudes, and dispositions that should be taught to young children.
8. *Everyday Learning About How Things Work* (2008) by M. Fleer (Early Childhood Australia). This book outlines how parents and teachers can use discovery and problem solving to help young children understand what goes on in their daily lives.

Children's Books

Cole, H. (2003). *On the Way to the Beach*. New York: HarperCollins/Greenwillow.

Dotlich, R. (2006). *What Is Science?* Macmillan.

Freeman, M. (2004). *You Are a Scientist*. Rourke Publishing.

Garrett, G. (2005). *Scientists Ask Questions*. Children's Press.

Lehn, B. (1999). *What Is a Scientist?* Brookfield, CT: Lerner Publishing Group.

Trumbauer, L. (2000). *Everyone Is a Scientist*. Maukato, MN: Pebble Books.

Websites

1. *Education.com* (http://www.education.com/activity/preschool/science/65) includes a section on Preschool Science Activities, including lesson plans and ideas for practical hands-on activities.

2. *The Amazing Preschool Activities* website (http://www.amazing-preschool-activities.com/preschool-science.html) is a wonderful resource for teachers.
3. The *U.S. Geological Service* website (http://education.usgs.gov/common/primary.htm) includes resources for teaching science in the primary grades, including teaching modules and classroom activities.

◇ Other Things You'll Need

- **Science Kits**

Using fanny packs or small backpacks, create a number of science kits for children's use. These kits, filled with science tools, can be available in the science area so that children who want to study their world can take them outside or use them to study something inside the room or school.

Stock each kit with

- a magnifying glass wrapped in soft cloth
- small plastic bags with twist ties to hold specimens or collections
- a tape measure
- small transparent boxes for collecting
- a small flashlight
- a small garden trowel or spoon for digging
- small pads of paper, pencil, and markers
- pieces of yarn

- **Clipboards**

Also have handy a number of clipboards with paper and markers. Plain newsprint is fine, or you could premark the paper (as shown here) to encourage children's data collections.

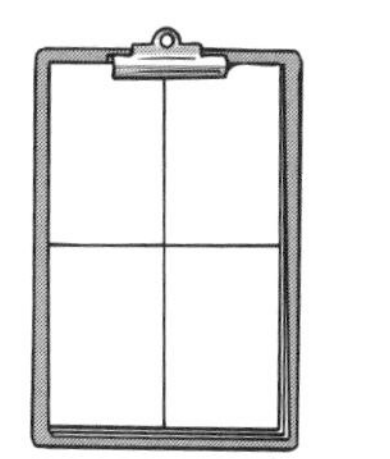
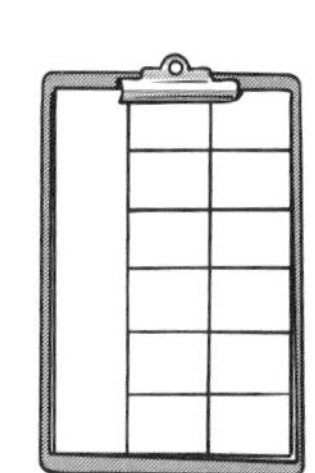
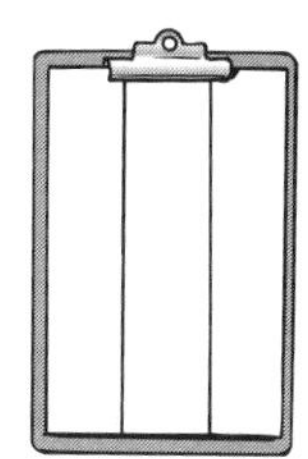

- **Scales**

Supply a balance scale and a regular scale.

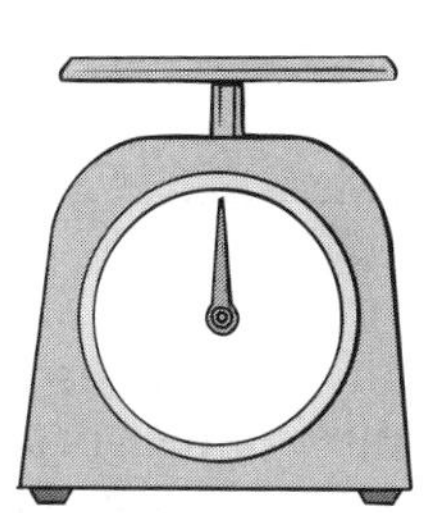

The Home–School Connection 

Children should develop the habit of using the scientific method at home as well as in school. Develop science kits for children to take home with them. Include a task card with each kit. The task card should explain how children use the kits at home. An example:

Task 1

- Send home a bag with an assortment of about 20 seeds; include different-colored seeds, different-shaped seeds, large seeds such as lima beans, and small seeds such as radish seeds. The kit should include paper, pencil, magnifying glass, small plastic containers, and cotton balls.
- Examine the seeds. How many different shaped seeds do you have? What colors are they? Are the seeds smooth, rough, round, or pointy? Draw a picture of the biggest seed. Use your magnifying glass to look at a small seed. Draw the small seed as you see it through the glass. Now draw the small seed as it looks without the glass.
- Put some cotton balls in one of your plastic bags. Now dampen the cotton with about 3 or 4 tablespoons of water. Put the two largest seeds on top of the damp cotton. Close the bag and leave it in a sunny spot, such as near your kitchen window. Observe what happened to the seed over the next 2 to 3 days. Draw a picture of your observations.

Task 2

- Send home a kit with an assortment of magnets, a small bag of paper clips, paper, and pencil.
- Explore your magnets. Find objects in your house that the magnet will attract. Draw a picture if the items that the magnet will attract.
- Find out how many paper clips the large magnet will attract. Carefully add one paper clip at a time until the magnet can hold no more. Record your result in your notebook. Next, try the same with the small magnet. Which magnet is the strongest?

◇ Documenting and Assessing Children's Learning

There are many ways to document children's learning as a result of their investigations. The teacher may want to ask children to reflect on their experiences and organize their ideas through charts, story displays, murals, and webs. Findings are simply arranged and easy to interpret when presented graphically. Using concrete found objects, children can document their findings as shown in the examples.

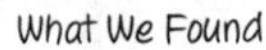
What We Found

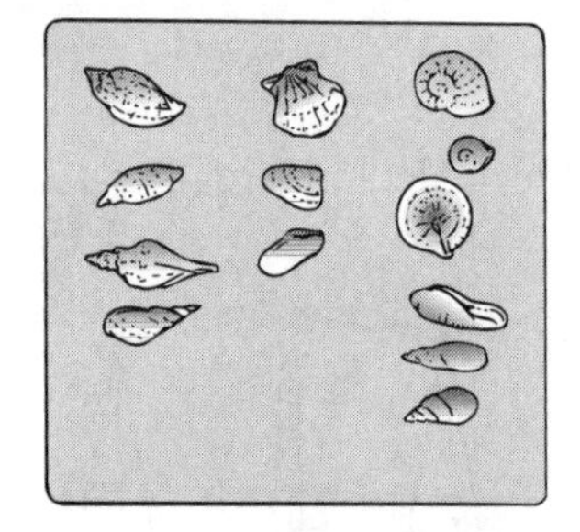

Seashells

There are many types of displays that children enjoy creating cooperatively. Cover a tabletop with a cloth or brown paper for children to display what they found on their walk around the school. Have each child contribute a sentence to a sign or chart describing what he or she did and found. For example, in the fall, Rachael said, "Leaves fall off the tree and fall on the ground." Kim dictated, "Once they fall on the ground, we can pick them up." Tyrone said, "Leaves are in different shapes and colors."

Take a walking field trip around the school to observe how many different kinds of plants you can find. As you walk, observe the children.

- What do they attend to as you walk along?
- How many questions are asked?
- Are children using scientific terms?
- What do they discuss?

To further evaluate children's growth in using scientific inquiry, observe when and how children use their science tool kits. Note when they are used, where, and how children use them. Are they doing more than just playing with the tools? Are they taking the leap into putting two or more ideas together to form generalizations? The tear-out sheets at the end of this chapter can be used at different times during the school year to chart growth in children's knowledge of science concepts and to help you plan your curriculum on the basis of children's current understanding. Chapter 7 has many other ideas for assessment and documentation.

FOR THE CHILDREN

1. Learning to Observe

◆ There are daily opportunities in the prekindergarten or kindergarten classroom to develop children's observation skills. Begin by including observation in the morning opening or calendar by asking children to observe the weather.

◆ Have students use their sense of touch to identify objects in a "mystery bag." For this activity, a number of objects are placed in a paper bag and—without looking inside the bag—the children feel the objects inside with their hands. They can then try to identify and name the objects.

Children learn to observe carefully.

- Another popular activity is the "secret cups" game. For this activity, you need eight or ten Styrofoam cups with lids. Place familiar objects inside each of the cups, such as a penny, some dried peas, liquid, feathers, and so on. All of the cups should then be sealed so that the children cannot see the objects. The children can then pick up the cups, shake them, and listen to the sound as they shake them. Their task is to identify the objects without looking inside. By changing the objects each week, this task could be an ongoing activity in the classroom.
- Activities involving texture are also popular with kindergarteners. Some of the tasks involving the sense of touch developed by Maria Montessori, for example, are still relevant today. One approach is to provide fabric swatches and have children identify them according to sight and touch. Then, after they become more proficient, they can wear a blindfold and rely solely on their sense of touch to identify the swatches of cloth. Similar activities can easily be designed and used to focus on other observable properties such as sound, color, texture, temperature, and weight.
- Construct "I spy" tubes using toilet paper or paper towel tubes. Make sure that you have enough for each child and provide an opportunity for the children to decorate their tubes. Show children how to use the tube as an "I spy" tube. Play the game "I spy." Look through your tube and say, "I spy something red and blue with white stars. What do I spy with my tube?" Have children use their tubes to find and name the thing you've spied. Once they get the idea of using a spy tube, ask children to lead the game.

2. Learning to Question

"Why is the sky blue?" "Who made the sky?" "How do birds fly?" Children ask questions constantly. Often they really do not want or need an answer; they simply are talking. With all the questioning children do, one might think they need no help in learning to ask questions. Yet children—like adults—need to be able to sense a problem and learn to ask and answer questions.

- Make sure that children have the psychological safety to feel free to question. Think about classes you have taken. When you feel insecure or think your question would sound silly to others, you won't ask it.
- Offer and provide knowledge for children to question. Sometimes you feel secure enough to ask a question but don't know enough to be able to frame or articulate the question.
    - Provide reference books on the topics children are studying or interested in.
    - Seriously discuss topics with children. Research shows that teachers spend more time giving directions or correcting children than actually talking with them about what they are doing or trying to learn.
    - Give children the vocabulary they need to ask questions. If you are studying butterflies, make a word hanger or chart with the words *chrysalis, cocoon, caterpillar,* and *butterfly;* if they are studying plants, the hanger might contain the words *leaf, bud, stem, sprout, seedling,* and so on.
- Teach children to pose questions. You might ask 3-year-olds to think about what they want to ask the police officer when he or she visits; 4- and 5-year-olds can be asked to think about what they want to ask while on a field trip. List the questions on a chart. Before going on the trip, cut the chart apart, and give each child or team of children one of the questions. Read the question to them to refresh their memories. Have the children ask their questions on the trip. Take along pads of paper or clipboards so they can record the answers.

3. Planning and Conducting Investigations

- Young children, who live from minute to minute, have difficulty thinking about the future. Thus, making plans is difficult for them. Still, you can introduce children to planning investigations.
  - Ask them what they will look for when they go on a nature walk around the block or what they think they will see. Use a KWL chart like this to begin and complete the study.
  - Have them make decisions about what they will take on the walk to collect things they find. They might elect to carry a plastic bag or a clear plastic box.
  - Ask children to decide how many tool kits and clipboards to take along and who will be responsible for these.
- Conducting investigations involves a number of methodologies. You'll want children to learn to practice skills of measuring, categorizing, and organizing their observations.
  - Measuring
    - Begin measuring with arbitrary measures. Children could find out how long and wide the sandbox is by walking the perimeter with their feet. Count and record the number of footsteps.
    - Walk different areas of the play yard to find out how big the yard or parts of the yard are. Again, count and record the number of footsteps.
    - Measure smaller things such as seeds, insects, or rocks. You might use small plastic blocks as the unit of measurement. Ask children to pick out the smallest, largest, widest, or thinnest insect.
    - Obtain a balance scale for children to use to find out which weighs the most:

      a large feather or a small marble

      a large empty box or a small unit block

      types of seashells

      acorns or pinecones
    - Talk about children's measurements. Ask children to predict which item will weigh the most.
  - Categorizing

    Set the stage for categorizing:
    - Children's natural need to find out and order the world in which they live leads them to categorize things in their world.
    - Make scrapbooks. Staple a couple of pages of paper between two sheets of construction paper and place the blank books in the writing area so that children can use them as they wish. Once in a while, set up a table for making specific scrapbooks. As the younger children cannot simultaneously handle magazines and cut out pictures, provide them with a bunch of precut pictures of animals, flowers, or whatever else you want children to categorize in their scrapbooks. Then they could make scrapbooks with themes of:

      Animals that Live in the Zoo, sorting zoo animals from precut pictures of farm, domestic, and other animals

      Birds I See, choosing birds they have seen from a variety of pictures of birds

Children's need to order their world leads them to categorize.

- Make categorizing trays. You could glue clear plastic boxes to a board. Then provide children with things to categorize, such as

  seashells

  nuts and bolts

  buttons too big to stuff in noses or other places

  bells—cow, decorative, sleigh, or other types

  fabric, wallpaper, or tile squares to sort

  large marbles of all types to play with and possibly begin sorting by color or type

  plastic animals to categorize as farm, zoo, pet, or wild

  old greeting cards to play with and sort

◇ Reflecting

What do scientists learn? What did we learn? These are the questions that might guide teachers in assisting children in the reflection process. As a scientist, each child has now participated in individual and group investigations. Questions have been recorded, and graphs and displays of evidence have been exhibited around the classroom with photographs taken by children and teachers. Children have recorded the processes that they followed in their science journals and in their labeled drawings.

Teachers can use previous documentation to draw children's attention to what they learned. There should be some culminating activity or activities so that the children can demonstrate their knowledge to others. At the end of their garden investigation, children at the Center for Young Children planned a party for parents. They formed committees on the basis of interest. The tasks of the decorations committee were

- Create a garden mural to hang outside.
- Make display boards describing the garden study to put in the classroom.
- Paint pictures for classroom frames.

The display boards documented the entire process that the children followed in their investigation and caused them to reflect on where they started (with a lot of questions) and where they finished (with some very good answers based on evidence). With older children, the teacher may go a step further and ask whether the investigation generated any further questions for study: "What would we still like to know about gardens?"

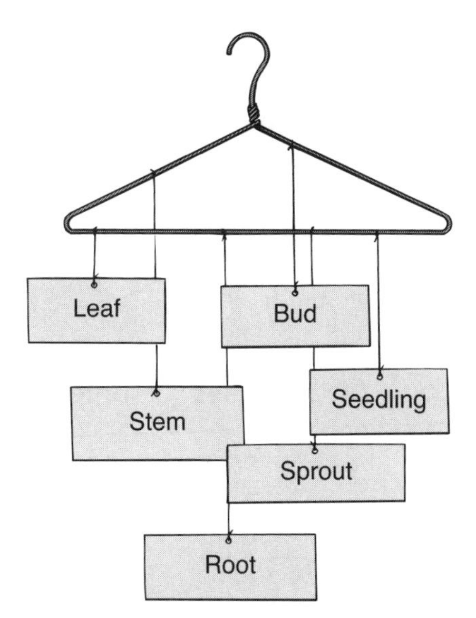

In conjunction with the party, children may want to pose scientific questions for their families to answer. They can provide tools and journals for families to record the process they followed and their conclusions.

◇ Extending and Expanding to the Early Primary Grades

Although children's concept development varies even within a single age group, older children have usually developed more abstract, critical thinking skills. Individual children are capable of conducting their own investigations after a discussion with the teacher.

Use Lisa Trumbauer's book *Everyone Is a Scientist* to inspire children to think as a scientist does:

- Discuss what things a scientist could study. (Of course, such a list could be endless.)
- Have children generate their list of things a scientist could study. Record the responses on a large chart. Record the questions that scientists would have about their study.
- Exhibit and discuss the tools of a scientist. *Everyone Is a Scientist* has photographs of an adult scientist and a child scientist at work.
- Have children generate a list of tools. Record these on the chart.
- Have children think about the appropriate tool to measure each topic that scientists could investigate (as recorded on their chart).
- Match the topic to the tool on the chart.
- Have each child pick a topic that could be investigated in the classroom or the play yard. For the children to conduct their investigation, they should choose appropriate tools and record their findings by writing in their science journals and taking pictures of things that they did to document their work. For example, if John investigated fall leaves, he might have recorded their shapes and colors and created a collage.

The teacher may want to create his or her own documentation by taking photographs of students using a tool, examining and measuring things, and recording the answers to the questions in their journals.

Individual Evaluation: Assessing Children's Science Skills

Date: ______________

Name: ______________

Age of Child: ________

	Always	Sometimes	Never
Observes various types of rocks	______	______	______
Identifies topics for investigation	______	______	______
Asks appropriate questions	______	______	______
Plans an investigation individually	______	______	______
Participates in group planning	______	______	______
Chooses appropriate tools to gather evidence	______	______	______
Exhibits good observation skills	______	______	______
Uses tools effectively	______	______	______
Gathers evidence appropriate to topic	______	______	______
Can articulate conclusions	______	______	______
Uses appropriate ways to document the process	______	______	______
Understands documentation (charts, graphs, and so forth)	______	______	______
Uses new scientific terminology	______	______	______

Comments:

7

Living Things Grow and Change: Seeds and Plants

FOR THE TEACHER

◇ What You'll Need to Know

Because young children are natural investigators, they are anxious to learn about the living things around them. Investigations into seeds and plants provide children with the foundation necessary to study the life cycle of animals and human beings. Developmental psychologists have found that children's understanding of the biological sciences develops gradually. The *National Science Education Standards* (NRC, 1996) note that young children have basic confusions about living and nonliving and tend to equate life with movement. Further, young children believe that all objects are "made for" a purpose.

Attempting to change children's understanding of the world through direct instruction does not seem to work (Seefeldt & Barbour, 1998). From a constructivist perspective, children are actively engaged in building theories about the world and the way it works. If teachers believe that children must actively construct their own knowledge, they will not *tell* them about science concepts. The teacher's role is to help children learn science through appropriate planning and through questions and comments that promote scientific thinking. "Inquiry into authentic questions generated from student experiences is the central strategy for teaching science" (National Academy of Sciences, 1995, p. 5). The key to effective science teaching is an emphasis on the process skills of observing, classifying, comparing, predicting, and communicating. Gradually, through active experiences with plant life, children will acquire the foundation on which later abstract learning can be built.

◇ Key Concepts

- Plants require air, water, food, and light to live.
- There are many kinds of plants, and each has its own form or structure.
- Plants make seeds.
- Seeds come in many different shapes colors and sizes.
- Seeds have three things in common: a protective seed cover, a baby plant (embryo), and a food supply.
- Seeds grow into plants with roots, stems, leaves, and flowers.
- Plants grow and change.
- Green plants respond to gravity, water, light, and touch.

◇ Goals and Objectives

The children will be able to

explore seeds and plants both outside and in the classroom;

identify the processes that permit plants to live;

observe, compare, and classify/measure a variety of seeds;

◆ Standards Alignment

Key Concepts and Goals and Objectives are based on the *National Science Education Standards* content standards, grades K–4. Standard C: Life Science (the characteristics of organisms, life cycles of organisms, organisms and environments).

Benchmarks for Science Literacy (The Living Environment)

observe, compare, and classify/measure a variety of plants;

match plants with the seeds that they make;

observe the growth of seeds into plants;

observe how plants change as they grow;

learn scientific terminology for their observations as appropriate.

◇ What You'll Need

You may want to refresh your memory about seeds and plants by consulting the "Outstanding Science Trade Books for Children," a list containing books published the previous year as a cooperative project between the NSTA and the Children's Book Council. The following resources offer information on plants and seeds:

1. *Plants on the Trail with Lewis and Clark* (2003) by Dorothy Hinshaw with photographs by William Munoz (Clarion). This book describes how Lewis and Clark painstakingly observed, collected, and catalogued plant life in their journey across the western United States.
2. *Autumn Leaves* (1998), written and illustrated with photographs by Ken Robbins (Scholastic). This book takes the reader on a walking tour of some of the most well-known autumn leaves and the trees from which they fall. The book features striking color photographs.
3. *Garden* (1998), written and illustrated with photographs by Robert Maass (Henry Holt). This book depicts the beauty and diversity of gardens. It includes information on the basic care of a garden through the seasons.
4. *National Audubon Society First Field Guide: Wild Flowers* (1998) by Anne Rockwell (Walker). This guide tells a story of what happens to one small bean when it interacts with some soil, a little water, sunlight, and a child's tender care.
5. *Our Wet World* (1998) by Sneed B. Collard (Charlesbridge). This book depicts the flora and fauna that inhabit the waterways of our planet.
6. *Buried Treasure: Roots and Tubers* (1998) by Meredith Sayles Hughes and Tom Hughes (Lerner). This book emphasizes food, observations, and simple experiments, which can be adapted for younger children.
7. *Peterson First Guide to Wildflowers of Northeastern and North-Central North America* (1998) by Roger T. Peterson (Houghton Mifflin) and *Peterson First Guide to Trees* (1998) by George A. Petrides (Houghton Mifflin). These are the first books that the beginning naturalist needs.
8. *Fostering a Sense of Wonder During the Early Childhood Years* (1993) by Ruth Wilson (Greyden). This is another good reference book for teachers.

Children's Books

Berenstain, S. (1996). *The Berenstain Bears Grow-It: Mother Nature Has Such a Green Thumb.* New York: Random House.

Bodach, V. (2007). *Seeds*. Minneapolis, MN: Capstone Press.

Carle, E. (1990). *The Tiny Seed.* New York: Simon & Schuster.

Carlstrom, N. (1989). *Wild, Wild Sunflower Child Anna.* New York: Macmillan.

Ehlert, L. (2005). *Leaf Man.* San Diego: Harcourt, Brace Jovanovich.

Florian, D. (1991). *Vegetable Garden.* San Diego: Harcourt Brace Jovanovich.

Fowler, A. (2001). *From Seed to Plant.* Danbury, CT: Grolier Publishing.

Graves, K. (1994). *Is It Alive?* Cypress, CA: Creative Teaching Press.

Hutts, A. (2007). *A Seed Is Sleepy.* San Francisco, Chronicle Books.

Iasevoli, B. (2007) *Plants!* New York: Harper Collins.

Kite, L. (1998). *Dandelion Adventures.* Minneapolis, MN: Lerner Publishing Group/Millbrook Press.

Raffi, (1997). *Everything Grows.* New York: Macmillan.

Richards, J. (2002). *A Fruit Is a Suitcase for Seeds.* Minneapolis, MN: Millbrook Press.

Rockwell, A. (1999). *One Bean.* New York: Walker Books for Young Readers.

Worth, B. (2001). *Oh Say Can You Seed? All About Flowering Plants.* New York: Random House.

Websites

1. *Backyard Plants* (http://www.backyardnature.net/botany.htm) contains an excellent collection of resources for teachers.
2. *Starting a School Garden* (http://aggie-horticulture.tamu.edu/kindergarden/Child/school/step.htm) contains step-by-step instructions for creating a school garden.
3. *Plants for Kids* (http://www.kathimitchell.com/plants.html) contains many links to other sites about plants and seeds.

◇ Other Things You'll Need

The following items will be useful when planning activities on seeds and plants:

- A variety of seeds (flower, fruit, and vegetable seeds; birdseed; seeds from trees such as fir cones, acorns, and chestnuts)
- Easily sprouting seeds such as black-eyed pea, lentil, lima bean, corn, pea, pumpkin, marigold, sunflower, and grass
- An assortment of fresh flowers and leaves and an assortment of dried flowers and leaves
- Nonliving things such as marbles, rocks, and stones
- Photographs and posters of flowers, plants, and trees at different periods of their seasonal cycle
- Large containers for planting and exhibitions
- Plastic or paper plates
- Magnifying glasses
- Soil or sand (indoors or out-of-doors)
- Watering cans
- Measuring tools
- Paper, glue, drawing tools, and other art supplies
- Access to the out-of-doors as a laboratory

Torquati and Barber (2005) have chosen some plants for easy gardening with young children:

- plants that attract butterflies: phlox, chocolate mint, dill, parsley, hollyhocks, milkweed, butterfly bushes, salvia, asters, trumpeter vine
- plants that provide color, texture, scent, taste: lamb's ears, sage, chives, mint, dianthus, sweet peas, zinnias, marigold, basil
- fruits and vegetables that are easy to cultivate: cherry tomatoes, yellow pear tomatoes, peas, string beans, melons, carrots, potatoes, beets, popcorn, sweet corn, broccoli

They also suggest that the teacher contact the local poison control agency to obtain a list of toxic plants that should never be permitted within the children's play yard.

The Home–School Connection 

There are many ways in which parents can assist children with science experiences about seeds and plants. First, they can encourage scientific thinking by asking open-ended questions and allowing time for children to formulate the answers. They can also foster the skills of science by inviting young eyes and fingers to notice small details as they take a walk through the park, helping children to put things in groups based on their characteristics, testing children's ideas about how the world works, and encouraging children to quantify the world around them (National Science Teachers Association, 1999).

On the specific topic of seeds and plants, parents may encourage children to explore their backyards or a neighboring lot or field for seeds, leaves, and other plant growth. Children may also make collections that would serve as the basis of a mini-museum. Teachers may also create science bags or backpacks to take home with simple experiments or suggestions for cooking experiences.

The U.S. Department of Education has published a book for parents by Nancy Paulu, *Helping Your Child Learn Science.* This small, parent-friendly volume is full of simple activities that parents can do with their children to enhance experiences with science. It employs a hands-on, discovery approach.

◇ Documenting and Assessing Children's Learning

Documenting and assessing children's concepts of growth and change in living things—seeds and plants—is a continuous activity done on an individual and group basis. Various forms of documentation may be used, including child self-reflections, observations of development (including developmental checklists and anecdotal notes taken at intervals by the teacher), individual portfolios (with examples of children's work, photographs, audio- and videotapes, structured interviews, children's self-evaluations), and products (group and individual). The teachers must be a keen observer and excellent collector. In addition, a camera is invaluable in the documentation and assessment process.

Teachers should begin the documentation process by assessing what children already know. At the Center for Young Children, as part of an investigation on gardening, the teacher made a large chart entitled, "What do you know about gardens and gardening?" Children's dictated responses filled the chart. For example, Daniel said, "You have to water them." His contribution was written down along with his name. After an investigation into the existing garden, the teacher made another chart entitled, "How can we make our class garden better?" Children's responses were recorded with their names as before. Many pictures were taken of the planting and growing process. Finally, when the garden was complete, a mural was created by the class to depict the garden and a party

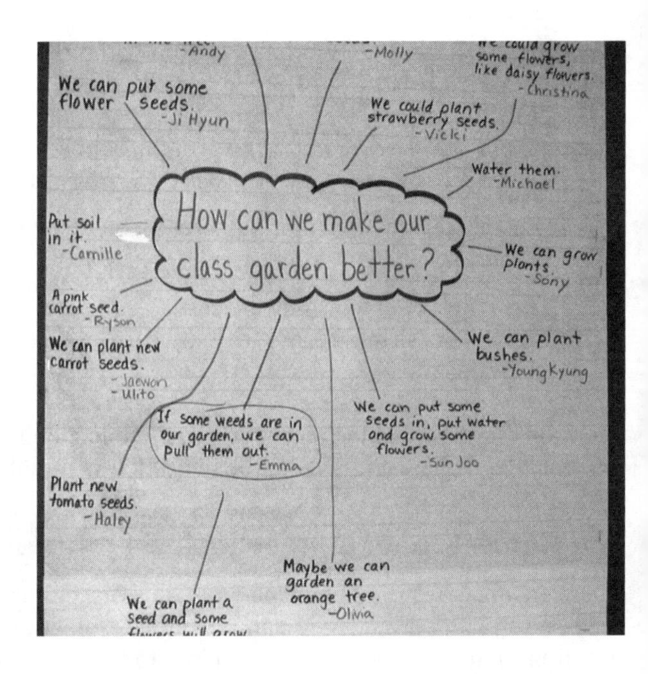

Children document each step of their investigation of gardens.

was held with exhibits of the children's work to convey their understanding to others. Care was given to the beauty and placement of exhibits so that children knew that their work was valued.

In assessing children's understandings of seeds and plants, teachers should collect a variety of evidence such as drawings, drawings with dictation, photographs, a record of children's command of science terms, and so forth. The evidence should be collected over time to gauge progress. For older children, science journals provide a good indication of children's thinking. Assessment is tied to goals and is always made to inform future instruction.

"Involving students in the assessment process does not diminish the responsibilities of the teacher—it increases them. It requires teachers to help students develop skills in self-reflection" (National Academy of Sciences, 1995, p. 10). The tear-out sheets at the end of this chapter can be used at different times during the school year to chart growth in children's knowledge of science concepts and help teachers plan a curriculum on the basis of children's current understanding.

FOR THE CHILDREN

◇ Investigations of Growth and Change with Seeds and Plants

1. Using Process Skills with Seeds

Children are attracted to all types of seeds. After a nature walk, what teacher or parent has not found a child's pocket stuffed with seeds that they wish to keep as prized possessions? Teachers should be sure that the seeds that children collect are not poisonous. Children enjoy the endless variety of shapes and colors and use seeds naturally to decorate their mud and sand creations as well as indoor art projects. Use their active experiences with seeds to build more formal concepts about plant life. Children should record all of their active experiences with science in a simple science journal.

Start with simple activities with younger children:

- Have children observe seeds carefully and classify them according to size, shape, and color. Remember that younger children may be able to classify by only one attribute. Provide children with magnifying glasses to use while they examine the seeds. Allow children to feel the seeds for hardness.

- Notice similarities and differences in seeds. For this activity, provide the children with several containers of different types of seeds, such as vegetable, fruit, and dried bean. Talk about how they are alike and different. Allow the children to take some seeds from each container and classify them. Some children will use color, some size, and some shape. If they cannot find a basis upon which to classify a seed, encourage them to figure out another category. Children may add seeds from snacks, from nature walks, and from home to the seed project.

2. Using Process Skills with Seeds and Plants

After experimenting with seeds of all types, children will be ready to begin to understand how plants grow from seeds and how living things grow and change. Teachers may want to begin using simple scientific terms at this point.

- Encourage children to open beans or corn seeds that have been soaked overnight to discover the baby plant or embryo. Using questioning techniques, ask children what they see inside the bean seed and what it looks like. Provide materials for children to discover whether the embryo will grow without the rest of the bean seed and which seeds, when soaked in water, will yield baby plants. Children may wish to draw pictures of the small baby plants and dictate stories about them.
- Watch to see if children will suggest planting seeds. Children can watch seeds sprouting by placing them on wet sponges. One class of 4-year-olds walked through the seeds outdoors with wet socks. The socks were then put on a large table outside to see which seeds would sprout on socks in outdoor conditions.
- Plant seeds as a small-group project with many choices for how children will undertake their own experiment. The teacher may want to have a discussion with the children first and chart the needed types of seeds, number of seeds, the type of soil, and the type of container. Children might want to decorate their containers to indicate the type of plant they predict will grow. Children should plant their own seeds. Although there may be some failures, they will help children to explain what happened. In one class, 3-year-olds learned that they could not forget to water their plants (even though the teacher, Ms. Gomez, had carefully put a small watering can next to the containers). The 4-year-olds in Mr. Green's class figured out that their seeds could not get air because they had been "buried." In most cases, it is important that children have success so that they will be able to describe what contributed to initial growth and how their plants grew and changed.
- Use two clear plastic tumblers to create a sprouting dome. Wet five cotton balls and use them to line the bottom of one of the tumblers. Then put two lima beans on the wet cotton. Upend the matching tumbler on the rim of the prepared tumbler. Tape the rims together to make a dome enclosure. Place the dome away from direct sunlight. Predict and record the growth of the lima beans.

Children enjoy growing flowers in the classroom.

- Make a plant grow toward the light. Plant one or two bean seeds in a small paper cup of soil and water as needed. Wait a few days until the leaves appear. Punch a small hole with the sharp end of a pencil in the side of a paper cup. Cover the growing plant with the paper cup and place in a well-lighted place. Uncover to water as needed. Make sure that the cup faces the light; after a few days, have the children observe what happened to the plant.

- Measure and chart change. Children can use measuring instruments to see how tall their plants are growing. Younger children can use nonstandard measures such as how many thumbs, paper clips, and so on to measure the growth. The different types of plants may be compared. Younger children could draw pictures to indicate the changes; older children might want to keep a record of the plants' growth over time.

- Keep a chart of new words learned and what elements were required to keep the plants alive. If the teacher allots time over the school year to the study of seeds and plants, it will be possible for children to observe the life cycle in action—children may raise beans, for example, and grow new plants from their bean seeds.

- Determine whether children have begun to understand the difference between "living" and "nonliving." For instance, Mr. Green put some marbles and rocks on a table with a few different types of seeds and plants. He asked the children to give him some ideas on how they could be classified. Then he made two columns and asked the children to classify on the basis of "living" and "nonliving."

- Build a terrarium. Help children create the conditions that will keep plants alive. You may want to include insects that children have captured on the way to school or at home. Children may observe and draw or chart their observations at various times of day or over the period of a week or a month.

- Build mini-museums to house seeds and plants that are alive or have been dried. This is an excellent project to undertake with parents (see the letter to parents in

Tear-Out Sheet 7-2 at the end of this chapter). Have a show at school to exhibit the mini-museums to children, families, and the community. As part of the program, invite parents to eat foods that have been created using the seeds and plants.

3. Using Children's Books to Motivate Active Experiences with Seeds and Plants

- Read *Dandelion Adventures* by L. Patricia Kite. Because the humble dandelion flowers almost anywhere, young children who live in urban environments will be especially interested in the journey of seven dandelion seed parachutes sailing through the air. Each land in a different place: garden, sidewalk crack, park, school yard, woodland, stream, and a faraway land. Most of the seeds produce new plants with a yellow flower that is pollinated by bees and produces new seed parachutes. The book ends with several pages of facts about dandelions. Have children find dandelions outside. Look for the seed parachutes and try to see where they land. Draw pictures or create a group book on the life cycle of the dandelion.

- *The Berenstain Bears Grow-It: Mother Nature Has Such a Green Thumb* by Stan Berenstain introduces children to simple plant science. There are simple instructions for planting seeds, cuttings, and tubers.

- Eric Carle's *The Tiny Seed* is another beautiful book that traces the growth of a seed to a flower. Carle's special collage technique should inspire the children to make a collage if the materials are present.

- Read Nancy Carlstrom's *Wild Wild Sunflower Child Anna* and take the children on a field trip right into a meadow like the one Anna, a small African American girl, enjoys as she talks to the flowers, whispers to the seeds, sifts soil through her fingers, and picks a yellow daisy. If there is no meadow available, try a park, an arboretum, or a greenhouse; or, like Tina in *Sunflowers for Tina* by Anne Baldwin, go on a site walk and discover a wonderful sunflower that has somehow sprung up on a nearby vacant lot. Be sure to take a camera on field experiences to record the children's discoveries.

- Douglas Florian's book *Vegetable Garden* helps young children learn how to harvest a vegetable garden. After the harvest, vegetables may be washed and eaten raw, cooked, or made into soup. Children can watch the transformation as they cook.

- Lois Ehlert's book *Leaf Man* inspires children to learn the names of various types of leaves. The beautiful drawings also motivate them to create artwork based on fall leaves and to use leaves they have collected to form shapes. Each page is cut and sized differently.

4. Using Arts and Crafts to Document Children's Active Experiences with Seeds and Plants

Plants and seeds lend themselves to beautiful artwork.

- As children explore in the play yard or park, encourage them to look carefully at the shape of leaves or the texture of bark. Although young children will not remember the names of the different trees that they see, they can compare and contrast the leaf shapes. Take paper so that children can make leaf or bark rubbings that they can share with other children. Make a chart comparing the shapes and textures. Exhibit the beautiful rubbings.

- Children could also make leaf prints from the leaves they have collected from the ground or iron (with your help) a leaf or an arrangement of leaves and seeds between two pieces of waxed paper.

- Create a root holder to view what goes on under the soil when the first roots of a baby plant are bursting out of the seed. See Tear-Out Sheet 7-6 for instructions on making a root viewer.

- Make seed castings by pressing dried pods and other seed heads with interesting shapes into flattened pieces of clay in any shape. The seed heads will probably fall out eventually, but children will be left with a lovely impression in the clay.

◇ Reflecting

Ask children to organize their experiences. Provide them with exhibit space so that they can display their collections, plants, and the mini-museums they have created with their parents. Provide poster paper for children to dictate explanations and directions to be hung above their work. Additionally, make sure that the children's science journals are up to date, illustrated, and displayed next to their experiments with plants and seeds.

Have a party! Invite family members and the school community to view the results of active science experiences. With the children, cook food that is made from plants and seeds. Have them observe how the ingredients change in the process of cooking. Serve the food to guests at your science party. (See Tear-Out Sheet 7-3 at the end of this chapter.)

◇ Extending and Expanding to the Early Primary Grades

Although the range of development is great within any group of children, the following experiences are appropriate for older kindergarteners and primary-age children in 1st through 3rd grade.

Children in the primary grades can do the following exercises:

- Expand their seed collections and sort them on the basis of more than one attribute, such as putting all the gray, hard seeds in one container and the soft, green seeds in another.

- Understand the process of photosynthesis: the concept that plants are the only things on earth that turn sunlight into food. Plants need sunlight, air, and water to remain healthy. Try these experiments from *Helping Your Child Learn Science* from the U.S. Department of Education (Paulu, 1992):
  - Have children look in a plant-care book or ask an adult to find out how much water each plant needs. Some may need to be watered more than others.
  - Have children take two clippings from one plant. Put one in a glass of water. Put the other one in a glass with no water. Check each day to see how long the one without water can survive.

- Have children observe and record what happens when a plant (or part of a plant) doesn't get any light. How long does it take for the plant to react? How long does it take for the plant to return to normal in the light?

◆ Ask better and more specific questions and make better predictions. They can also draw or record their conclusions in writing. For example, using the previous experiments or others, have children make predictions about what will happen under different conditions before they observe and record the results.

◆ Collect and preserve plant specimens with the help of an adult who knows which are poisonous and which are rare and not to be collected. Children can make a herbarium by mounting pressed plants and recording the name of each plant that they are able to identify.

◆ Understand habitats. Help the children construct an insect terrarium. Some insects, such as grasshoppers and crickets, may be kept in a terrarium with a gauze or screen top. Plants and animals that live together in similar habitats out-of-doors should be selected. Air and drainage are important. A bowl of water with stones or a twig over it completes the terrarium. Children may then make continuous observations and predictions about how the insects and plants will behave.

◆ Children can design their own simple experiments. Have children generate questions about plants, seeds, and insects. Then have them design an experiment that will answer their questions. With your assistance, they may consult reference books. For example, a child might want to know if ants eat the spilled food on the sidewalk on the spot or carry it back to their anthill. Find out the answer through active experimentation.

◆ Profit from a field trip to a hands-on science museum, an arboretum, or the laboratory of a university. Preview the site ahead of time and make sure that the personnel know how young children learn. Introduce the trip and ask the children to generate questions to be answered. Upon your return to the classroom, reflect upon the learning and document each new bit of information by making charts, posters, or books about the experience.

◆ Use computers to locate the websites of science museums and other resources for children. Teachers should take a preview field trip to make sure that the sites are suitable for the children.

◆ Build a larger vocabulary of scientific terms. These may be recorded in their science journals and posted next to experiments for easy reference.

[Date]

Dear Parents,

Our class is beginning a new science unit on the theme **"Living Things Grow and Change."** The children will be experimenting by planting different types of seeds and observing how plants grow. We will also be designing, planting, and caring for our very own class garden. This will be an active and busy project for the children, so your help in the classroom would be appreciated. Please let me know in advance if you are available to assist in any way with this project.

We hope that you will be able to volunteer in the classroom. If you are not able to visit our classroom, there are many things you can do with your children at home to reinforce the science concepts they have been learning about in school. Here are some suggested activities for you and your child:

- Talk with your child about what you see outside in the natural world.
- Help your child make collections of seeds and leaves.
- Go for a walk in the park and see how many different plants you can find.
- Visit the store and count how many different types of vegetables you see.
- Visit the flower shop to observe all the different flowers.
- Help your child use reference books or websites to find answers to questions about how things grow and change.

Your children will be recording their observations in a science journal. When they bring the journal home, please help them illustrate their observations. They will also need your help in labeling their pictures and perhaps dictating a sentence or two for you to write down about plants and seeds.

Thank you for helping with our new science unit. If you have any questions or comments, please drop me a note.

Sincerely,

[Signature]

[Date]

Dear Parents,

In the next few days, I will be sending a small backpack home with your children that contains suggestions for things they can do at home to learn about science concepts. These fun activities include:

- Planting seeds
- Using seeds, plants, and flowers to create a collage
- Cooking with seeds and plants

The science backpack will contain the materials as well as simple directions. Also included will be directions for creating a mini-museum with your child. For this activity, all you need is a shoebox or cardboard box, seeds, plants, dried flowers, glue, and art materials. The directions will guide you and your child in creating something beautiful to bring to school.

We hope that you and your child will enjoy these activities. Please don't be too concerned if the projects aren't perfect. Remember that children also learn when they make mistakes. I look forward to hearing about your projects. Please call, visit, or send a note to let me know how the projects are going. If you prefer, you can just send me a photo of your project.

Finally, please encourage your children to use their science journals to record their questions, observations, and the results of the project. They will enjoy drawing pictures in the journal, but they will need your help with the writing.

Thank you for working with your child on this project and for your continued support. I will be in touch in the next few weeks with information about a class party and exhibit to conclude our science unit **"Living Things Grow and Change: Seeds and Plants."**

Sincerely,

[Signature]

Science Exhibit and Class Party

[Date]

Dear Parents,

We are about to conclude our science unit **"Living Things Grow and Change: Seeds and Plants."** I hope you have enjoyed the activities that you have done at home with your child. I know that your children have enjoyed all of the activities. Your participation and support helped the children learn many new science concepts. I really appreciate your support with this project.

Now it is time to have a party to recognize all of the hard work that you and your children have done in the area of science. You and your family are invited to our party at _________ on _________. We will have child care for younger children. Please let me know if you need help with transportation.

The class party will also include exhibits of all the beautiful mini-museums that you and your children created. Also on display will be the children's science experiments, artwork with plants and seeds, and posters of science words the children have learned, as well as other science posters.

The children have been cooking with plants and seeds. They are surprised that some of the food is familiar to them. Other things are new to some children, such as toasted pumpkin seeds. At the party, we will have plenty of food on hand for all of us to sample.

Once again, thank you for your continued support, and I look forward to seeing you at the class party.

Sincerely,

[Signature]

Group Observation: Science Terms

Date: ______________________

Center/Area: ________________

Children's Names: __________________

	Accuracy		
Science Terms Used	**Not at All**	**Some**	**Accurate**
1. baby plant or embryo	_____	_____	_____
2. sprouting	_____	_____	_____
3. terrarium	_____	_____	_____
4. seed parachutes	_____	_____	_____
5. texture	_____	_____	_____
6. photosynthesis	_____	_____	_____
7. poisonous	_____	_____	_____
8. habitat	_____	_____	_____
9. classify	_____	_____	_____
10. growth	_____	_____	_____

Comments:

Individual Evaluation: Assessing Children's Science Skills
Living Things Grow and Change: Seeds and Plants

Date: ________________

Name: ________________

DOB/Age: ______________

	Always	Sometimes	Never
Explores seeds and plants	_____	_____	_____
Identifies the processes that permit plants to live	_____	_____	_____
Observes a variety of seeds	_____	_____	_____
Compares a variety of seeds	_____	_____	_____
Classifies and measures a variety of seeds	_____	_____	_____
Observes a variety of plants	_____	_____	_____
Compares a variety of plants	_____	_____	_____
Classifies and measures a variety of plants	_____	_____	_____
Matches plants with the seeds they make	_____	_____	_____
Observes the growth of seeds into plants	_____	_____	_____
Observes how plants change as they grow	_____	_____	_____
Asks appropriate questions	_____	_____	_____
Reaches appropriate conclusions	_____	_____	_____
Uses appropriate terminology (age level)	_____	_____	_____
Can apply concepts from children's literature on growth and change	_____	_____	_____
Enjoys constructing simple experiments (with the help of a teacher or parent)	_____	_____	_____
Enjoys using plants and seeds in artwork	_____	_____	_____

Notes/Comments:

Instructions for Making a Root Viewer

1. Take a 1-quart, square-bottomed, freezer container.
2. Cut from a top corner straight down to the middle of the bottom of the same side.
3. Make the same cut on the opposite side of the container.
4. Make a straight cut across the bottom.
5. Remove the cut-out piece of plastic.
6. Using plastic tape, attach a piece of Plexiglas to eh open side of the container. Make sure there are no holes or gaps.
7. Tape a piece of black paper to the top edge of the Plexiglas so that it can be lifted and lowered.
8. Place stones in the bottom of the container for drainage.
9. Add potting soil to fill.
10. Push a few seeds into the soil about one-half inch back from the Plexiglas edge.
11. Flip the black paper down.
12. Water the seeds and place the container in a dark spot until the seeds sprout.
13. Place the container in the light and lift up the black paper every day to check on the roots.

8

Living Things Grow and Change: Insects and Small Animals

FOR THE TEACHER

◇ What You'll Need to Know

Young children are highly interested in living things and are usually able to identify those they have an interest in or have had the opportunity to observe through books, films, or—best of all—directly. Most children ages 3 to 5 can describe the attributes of insects and small animals and distinguish one from another. They are familiar with bugs, although they may not know the proper terms to describe them. They have usually been exposed to fish, birds, some wildlife, and pets that live in their homes or in the classroom. It is wise to keep in mind, however, that some children are afraid of animals and need more support in their explorations.

The teacher should provide opportunities for children—especially those who live in circumstances that limit their interaction with small animals—to observe a variety of animals in the classroom, on the school grounds, in the neighborhood, at home, in parks and gardens, and at the zoo. Yet observing is not enough. As a teacher, you should encourage young children to ask questions to which they can find answers by looking carefully at small animals and then checking their observations with simple field guides and with one another. Young children need to experience science. As with seeds and plants, the key is an emphasis on the process skills of observing, classifying, comparing, predicting, and communicating findings. Gradually, through active experiences with small animals, children build concepts that will enable them to understand the natural world and move toward concepts of human development. It is necessary to limit experiences for young children to a few organisms because there are thousands of different species of insects.

Children at the Center for Young Children at the University of Maryland conducted an investigation of rabbits. First, teachers determined what children already knew about rabbits. They documented these findings on a big chart. Then, based on that information, a number of experiences were selected as part of the study. Children determined what kind of rabbit food they liked, such as lettuce, apples, Cheerios, peas, and carrots. Children's pictures were placed underneath each of the headings. Using Venn diagrams, rabbits were compared to guinea pigs, hermit crabs, and fish. Correct terminology was used as much as possible. For example, children said that hermit crabs had "antennas" and "pinchers." Finally, children in groups wrote rabbit stories, which were typed and beautifully exhibited with the children's pictures beneath them.

◇ Key Concepts

- There are many kinds of animals.
- Some animals are alike in the way they look and in the things they do, and others are very different from each other.
- Animals need air, water, and food.
- Animals can survive only in environments in which their needs can be met.
- Many animals make shelters to rear their young.
- Stories sometimes give animals attributes that they really do not have.
- Animals have life cycles that include being born, developing into adults, reproducing, and dying.

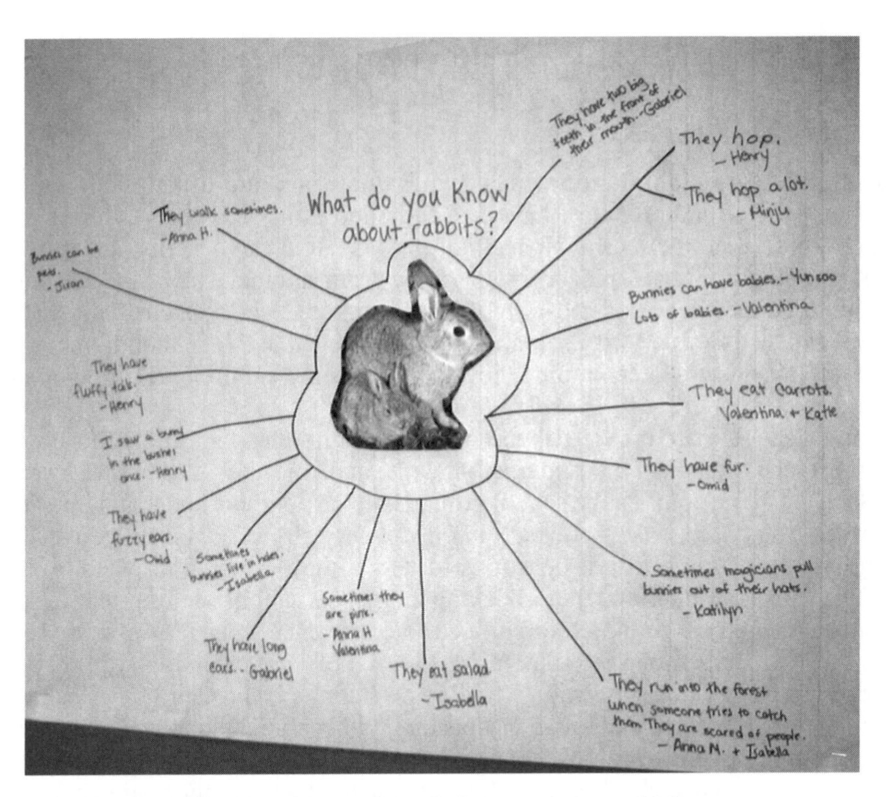

Teachers determine what children already know about rabbits.

> ### ◆ Standards Alignment
>
> Key Concepts and Goals and Objectives are based on the *National Science Education Standards* content standards, grades K–4. Standard C: Life Science (the characteristics of organisms, life cycles of organisms, organisms and environments).
>
> *Benchmarks for Science Literacy* (The Living Environment)

◇ Goals and Objectives

The children will be able to

- observe small animals, both outside and in the classroom;
- compare various insects and small animals and identify the similarities and differences;
- learn simple scientific terminology for animal parts and their functions;
- observe the life cycle of a small animal;
- identify environments where various small animals thrive;
- identify various shelters that small animals create for their young (nests);
- identify foods that are specific to different animals;
- identify stories in which animals are depicted with human attributes and feelings.

◇ What You'll Need

As for all science experiences for young children, you may want to consult the list "Outstanding Science Trade Books for Children" published by the NSTA and the Children's Book Council. This list is available on the NSTA website at http://www.nsta.org/publications/ostb/.

The following books, which are primarily for adults and older children, have resources on small animals:

1. *About Arachnids: A Guide for Children* (2003) by Cathryn Sill (Peachtree). Beautiful, full-page color plates in this book enhance a simple discussion of arachnids, which include spiders, scorpions, and tarantulas.
2. *A Place for Birds* (2009) by Melissa Stewart (Peachtree). This book introduces readers to the birds of North and Central America. It includes maps, project ideas, and ideas to protect the ecosystem.
3. *Brilliant Bees* (2003) by Linda Glaser (Millbrook Press). This simple and informative text contains information about the importance of bees to the ecosystem.
4. *Up, Up, and Away* (2009) by Ginger Wadsworth. This book follows the life of a spider as it grows from a baby to an adult. It includes beautiful watercolor drawings.
5. *How Birds Build Their Amazing Homes (Animal Architects)* (1998) by W. W. Robinson (Black Birch Marketing). Each chapter introduces a specific nest type (e.g., sewn, mound, clay), and the book has excellent photographs.
6. *National Audubon Society First Field Guide to Mammals* (1998) by J. Grassy and C. Keene (National Audubon Society). With the aid of the teacher, this book can be used with young children to verify observations.
7. *Insects (National Audubon Society First Field Guide)* (1998) by C. Wilston (National Audubon Society). As with the previous field guide, with the aid of the teacher, this book can be used with young children to verify observations.
8. *Animals in the Classroom* (1998) by D. Kramer (Addison-Wesley). This book has good information on the correct handling of classroom animals.
9. *Creepy Crawlies and the Scientific Method* (1998) by S. Kneidel (Fulcrum). This is an excellent guide for the teacher on setting up experimental conditions for children's scientific observations.
10. *Spinning Spiders* (2003) by Melvin Berger (HarperCollins). This book is part of the HarperCollins Let's-Read-and-Find-Out Series of Science. It includes activities and websites.
11. *Eliza and the Dragonfly* (2004) by Susie Caldwell Rinehart (Dawn Publications). This book tells the story of Eliza, her entomologist aunt, and the life cycle of a dragonfly.
12. *Lizards* (2003) by Daniel Greenberg (Marshall Cavendish). This book allows students to learn in depth about lizards.
13. *Wonderful Worms* (1992) by Linda Glaser (First Avenue Editions). This book contains simple yet colorful drawings and includes great facts to introduce worms to your students.
14. *Face to Face with Animals* (2007) by Darlene A. Murawski (National Geographic Books). This book focuses on observations made from nature photographs. It includes information on animals and how they live. It also teaches children that they can document through photographs as well as learn ways to help the animals they come face to face with.

Children's Books

Asch, F. (1999). *Baby Bird's First Nest.* San Diego, CA: Gulliver Books.

Carle, E. (1990). *The Very Quiet Cricket.* New York: Philomel Books.

Carle, E. (2004). *Mister Seahorse.* New York: Philomel Books.

Dorros, A. (1988). *Ant Cities (Let's Read and Find Out).* New York: HarperTrophy.

Fleming, D. (2007). *In the Small, Small Pond*. New York: Henry Holt.

Glaser, L. (1994). *Wonderful Worms.* Brookfield, CT: Millbrook Press.

Gracia, K. (2004). *Little Bug & Friends. It's Good to Be Small*. Bloomington, IN: Authorhouse.

Green, J. (2002). *In a Backyard*. New York: Crabtree Publishing.

Heiligman, D. (1996). *From Caterpillar to Butterfly (Let's Read and Find Out).* New York: HarperTrophy.

Laden, N. (2000). *Roberto, the Insect Architect.* San Francisco: Chronicle Books.

Llewellyn, C. (2005). *The Best Book of Bugs*. London: Kingfisher Publishing.

Palotta, J. (1986). *The Icky Bug Book.* Watertown, MA: Charlesbridge.

Parker, N. W., & Wright, J. R. (1987). *Bugs.* New York: Scholastic.

Rockwell, A. (2001). *Bugs Are Insects*. Harper Collins Publishing.

Rogers, F. (1988). *Mister Rogers' First Experience Book: When a Pet Dies.* New York: Putnam.

Sandved, K. (1999). *The Butterfly Alphabet.* New York: Scholastic.

Sayer, A. (2005). *Ant, Ant, Ant! An Insect Chant*. Lanham, MD: Northword Books.

Scarborough, K. (1997). *Spider's Nest (Watch It Grow).* New York: Time Life.

Simon, S. (2002). *Baby Animals.* New York: Seastar Books.

Starosta, P. (2005). *The Bee: Friends of the Flowers*. Watertown, MA: Charlesbridge.

Swope, S. (2004). *Gotta Go! Gotta Go!* New York: Farrar, Straus, and Giroux.

Voake, S., & Voake, C. (2010). *Insect Detective.* Somerville, MA: Candlewick Press.

Walsh, E. S. (2003). *Dot & Jabber and the Big Bug Mystery.* San Diego, CA: Harcourt Children's Books.

Zakowski, C. (1997). *The Insect Book: A Basic Guide to the Collection and Care of Common Insects for Young Children.* Highland City, FL: Rainbow Books.

Websites

1. *Pestworld for Kids* (http://www.pestworldforkids.org/guide.html) includes information, games, activities, and lesson plans about a whole range of insects and bugs, including bees, beetles, flies, mites, moths, and wasps.
2. *Insects.org* (http://www.insects.org) is an excellent site for teachers to learn about insects.
3. *Sialis* is (http://www.sialis.org/index.html) a resource for people interested in helping bluebirds and other native cavity-nesters survive and thrive. The website includes information about a variety of birds as well as photos of their nests.

◇ Other Things You'll Need

- Temporary cages for insects
- Live insects
- Preserved insects

- Live mealworms
- Live earthworms
- Hand lens and magnifying glasses
- Paper
- Crayons and markers
- Garden trowels and a digging place
- Dishpan
- Moist cornmeal or bread crumbs
- Large clean jars
- Fresh leaves
- Waxed paper
- Rubber bands
- Paper bags
- Notebooks or science journals
- Sketch pad
- Net for catching insects
- Plastic or metal cages for classroom animals
- Access to the out-of-doors
- Measuring tools
- Mounted pictures and photographs of various types of small animals

The Home–School Connection

As with seeds and plants, questioning, listening, observing, and predicting work best when attempting to convey concepts about animals to young children. Although often messy and time-consuming, the hands-on approach fosters concept learning. Some families have small animals in their homes. Small dogs, cats, guinea pigs, and gerbils provide an excellent opportunity for children not only observe, formulate, and answer questions about animal behavior, but also to learn good practices for caring for their pets. Because many families choose not to have pets, parents may encourage their children to search their homes and neighborhoods for observer-friendly small animals.

Some bugs are not "observer-friendly," so parents should set clear guidelines for children on what to look for. Ants, almost all spiders, fleas, silverfish, moths, flies, and ladybugs are harmless. Parents can encourage children to observe how some insects such as ants work as a community. Children can watch ants around their anthills and observe how they "notify" other ants of food. In the process of observations, parents may pose questions and encourage children to dictate or write the answers in a science journal. Drawing pictures will assist children to remember their observations. *Helping Your Child Learn Science* by Nancy Paulu (1992) is a helpful book for parents.

◇ Documenting and Assessing Children's Learning

Documenting and assessing children's concepts of growth and change in living things—insects and small animals—is a continuous process done with individuals and groups.

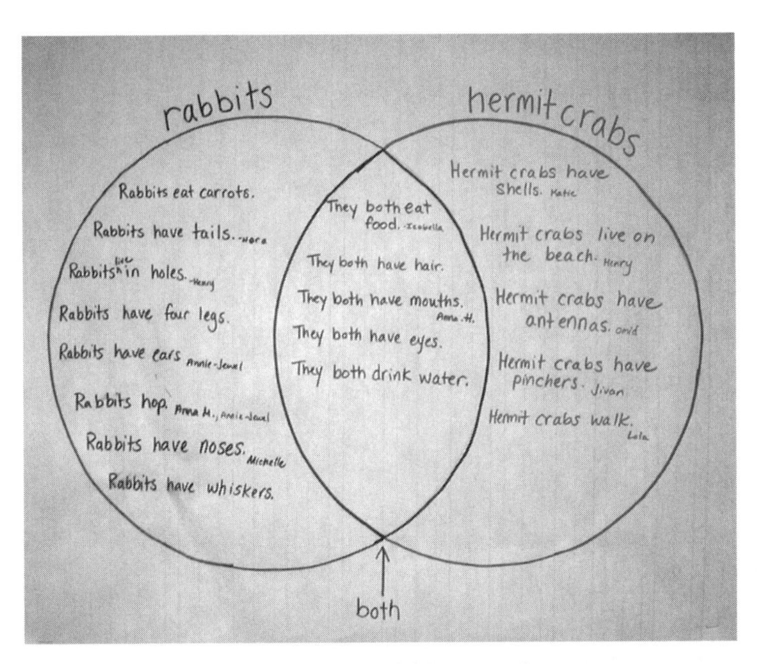

Using Venn diagrams, children compare rabbits to other creatures.

Chapter 7 provides a thorough review of documentation and assessment with young children. Teachers can observe what children are learning in the group context as children discuss and sometimes dispute others' findings. Various forms of documentation may be used. Children's self-reflections on their work may be recorded and displayed. The teacher should take anecdotal notes based on observations on the concept development of individual children and groups at intervals during an investigation. For example, what does Danny know now that he didn't know before as a result of comparing rabbits with hermit crabs or tasting rabbit food? What scientific term is he using that he was not using at the outset of the investigation?

The teacher should create individual portfolios for each child. Portfolios vary, but they usually contain examples of children's work over time, photographs, audio- and videotapes, or structured interviews in which children are asked questions directly. The checklist Tear-Out Sheet 8-4 at the end of this chapter will be useful to the teacher for documenting and assessing children's work. Children's interest level, products, and dictated stories about animal life all provide evidence to the teacher of the success of the investigation. For older children, up-to-date science journals should be examined and discussed with the child.

Tear-Out Sheets 8-3 and 8-4 at the end of this chapter can be used at different times in the school year to chart growth in children's knowledge of insects and small animals and to help plan a curriculum to extend children's current understanding.

FOR THE CHILDREN

First, the teacher should introduce the topic of animals by arranging books, posters, and animal exhibits around the classroom. A discussion may follow in which children generate a list of animals based on the definition that "an animal is any living thing that is not a plant." Then a decision must be made by either the teacher or the children (with the teacher's guidance) about which animals to study. Use children's books as motivators.

1. Observing Insects

Children should have no difficulty locating creatures under rocks, on window frames, and on weeds. Catching them is more of a problem and should be done in small groups with care taken to not harm the insect. Open plastic jars, sandwich bags, and nets make good temporary places to house insects. Tongs or Popsicle sticks may be used to capture and transfer them to a more permanent cage for observation. A simple temporary cage for small insects can be made by covering a plastic container with a piece of nylon hosiery stretched and held in place with tape or a large rubber band. Respect for the environment is taught when the teacher asks children to release insects at the end of the day. One or two interesting insects may be preserved for displays.

Start with simple activities:

- Suggest that the students look for the things that identify an animal as a member of the insect family: three body parts, six legs, and two feelers.
- Encourage children to draw pictures of the insect they are observing and dictate a story or label insect parts. If the drawings appear immature, encourage them to look carefully and enhance the detail in their drawings.
- Encourage the use of proper terminology by using words such as *thorax* and *antennae.* These may be recorded on a chart as they come up in discussions.
- Compare and contrast insects by the way they look, the things they do, and the foods they eat. Have children make charts or a Venn diagram listing similarities and differences.
- Observe insects' social behavior. For example, fireflies may "light up" to attract mates.
- Observe how insects defend themselves. Some have coats of armor; others sting, like bees and wasps.

Ms. Johnson, a teacher of 4-year-olds, posed this question to her class: "Is a spider an insect?" The children immediately responded that it was. She then asked them how they could be sure that they were right. They knew that they had to observe a spider and some of the insects they had collected. Small groups went to observe, recorded their conclusions, and shared them with the entire class. The answer: A spider is not an insect. Encourage children to pose their own questions and undertake investigations to answer them.

2. Observing Earthworms

The best place to find earthworms is in rich garden soil. Either dig them up or wait until after heavy rain, when the worms come to the surface at night. Try to collect the large night crawlers. If you prefer, you can purchase earthworms (also called night crawlers) from the fishing and hunting department of your local large retail store. Keep the worms in a box containing equal parts of peat moss and rich garden soil. Cover the box with a damp piece of cloth towel. Keep the earth and towel damp but not wet at all times. About once a week, feed the worms a quarter cup of oatmeal or bread that has been softened in water.

- Place an earthworm on a damp paper towel. Have the students observe how it crawls by lengthening and then contracting its body. Help the children find the head of the earthworm by watching closely for the lip or pharynx that the worm pushes out as it moves along. Touch the head and note how the worm contracts the front part of its body—it may even crawl backward.

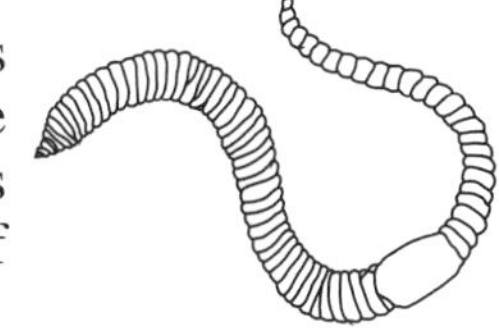

- Have the children pass their fingertips gently along the underside of the earthworm and feel the bristle (setae). Observe these feet through a magnifying glass. See what happens when you place the earthworm on a damp, smooth glass surface, where there is nothing for these bristle feet to grip.

- Place an earthworm on piece of damp paper towel and cover it with a saucer so that it cannot escape. Keep the worm in a dark room for about an hour, then shine a flashlight on the worm's head. Note how it quickly pulls its head away from the bright light.

- Put an earthworm on a rough paper towel and then on a smooth surface. On which does it travel easier?

- Present discovery problems to the children such as: "In what ways can you tell different earthworms apart?" "Which food do they seem to like best?" "Do earthworms like moisture?" or "Will an earthworm move to a dark or light place?"

3. Environment and Life Cycles

Mealworms are easy to maintain in the classroom and may be purchased cheaply at pet stores, eliminating the problem of catching them. They must be put in plastic or glass containers because they can chew through cardboard and Styrofoam. They should be placed in bran and covered with a moist paper towel. Very little water is required because they are able to extract moisture from their food. Children may feed them small pieces of potato, apple, or broccoli. At this point, explain to the children that in order to grow healthy mealworm beetles, they need to create a growth environment or habitat just as the mealworm would do for itself.

- Have children take time to observe each day what is happening in the mealworm environment.

- Let students record observations in their journals with either descriptions or pictures.

- Have students watch as the life cycle replicates itself as the larvae seem to "die." Actually, they are beginning a chrysalis stage. They will emerge from this as light-colored beetles about an inch long. They will mate and die in about a week. Their almost-invisible eggs will repeat the process, however. This is an easy way to observe the stages of metamorphosis, and the teacher may want to compare it to the caterpillar (see the Children's Books section of this chapter). *Gotta Go! Gotta Go!* by Sam Swope describes how a creepy crawly little bug fulfills her destiny to become a beautiful monarch butterfly by joining other migrating butterflies in Mexico. Deborah Heiligman's book *From Caterpillar to Butterfly* treats the topic in a more serious way. Have children measure and chart the changes in the mealworms that occur over time.

- Observe caterpillars. Remember that all a caterpillar does is eat. Caterpillars collected in the spring will finish their cycle in a month or so. Fall caterpillars will not hatch out of the pupa stage until the following spring. Make sure to add a stick to your jar and keep a fresh supply of leaves until the caterpillars have gone into their pupa stage. Leave a wet cotton ball in the jar. They need moisture in order to hatch. After your butterfly or moth has hatched, let it go.

- Have children dictate new vocabulary words that emerge from their life-cycle observations.

- Introduce children to a sea environment through Eric Carle's beautiful book *Mister Seahorse.* Explain to children that in most fish families, after the mother has laid the eggs and the father has fertilized them, the eggs are left on their own. But there are exceptions, and one is the seahorse. The female seahorse lays her eggs into a pouch on the male seahorse's belly, where they remain until they are fully formed and ready to emerge. Children should enjoy the idea that the "caring" parent is the father. Children can also learn about camouflage because acetate pages mask various creatures as Mr. Seahorse goes by. By removing the acetate page, children can view the fish that is hiding.

4. Nests

Most children enjoy watching birds. One way to be sure to get a good look at birds is to set up a feeding station with a birdbath. Children can observe the various foods that birds eat, such as sunflower seeds, chopped nuts, peanut butter, and fruit. By using birds as an example, teachers may set up experiences comparing foods that are specific to different animals: "How is bird food different from the food we feed our classroom fish?" Help children identify shelters that birds have created for the nurturing of their young.

- Help the children create simple homemade bird feeders. Try a pinecone that has been rolled in peanut butter. String Cheerios through a piece of yarn. Fill half an orange or grapefruit with bird seed.
- Have children try to spot a bird carrying grass or a twig in its beak and watch to see where it goes. Have children keep track of the nest over the season to see how the adult bird cares for its young.
- Start looking for active nests in the spring. Some birds nest under eaves and ledges and in other protected areas.
- Instruct the children to move slowly and quietly when they are near the nest because some nesting birds are easily disturbed by people.
- Use simple binoculars to observe birds at a safe distance.
- Start looking for abandoned nests to study in late fall. Ask children to identify the nest materials that were used by inspecting the surface of a nest.
- Have small groups of children take the nest apart bit by bit. Have them sort the materials into piles and see if there are any that they recognize.
- Have children check their observations with field guides and simple charts.
- Use science journals to draw pictures or record observations.

Note: In the United States, it is against federal regulations to collect the nest of any migratory bird or endangered species. If you are in doubt, contact the U.S. Fish and Wildlife Service.

5. Recognizing Children's Books in which Animals Are Depicted with Human Attributes and Feelings

Most of the children's books suggested earlier do not portray animals as people who talk and have feelings. The Eric Carle books are an exception, but they are strongly based on fact. Make sure that children experience these books so that they have an accurate view of animal science. You may want to read a book in which wild animals are given names, making them seem more like pets than part of the natural world. Sometimes motives are attributed to animals that make them seem human: "The fox wondered when the snow would melt so he could find food to eat." After reading the different types of books, ask children

- how the books are different, and
- which books seem to portray animals like the ones they have been studying.

Teachers can also read stories in which animals have human roles, but it is important for children to understand that these are fantasy and that they could not really happen in the world as we know it. Children may not master this concept until they are in the middle primary grades, but it is essential to their understanding and respect for the animal world.

6. Classroom Pets or Not

Teachers need to examine their feelings about having pets reside in the classroom. Too often, classroom pets live under poor conditions. Children have not been instructed in the proper handling of them, and there is nowhere for them to go on weekends and holidays. J. Davis (2001) tells of the near death of a beloved classroom pet when their Berkshire rat Pork Chop was dropped by a child. Pork Chop survived, but his cage mate, Curious George, "was unintentionally loved to death" (p. 85). These are poor messages to send to young children, both about death and about the dignity of life. In addition, if pets are to be a part of the science learning environment, children should observe and compare them, predict their behavior, test predictions, and communicate about them each day. Very often, pets get stuck in the corner of the classroom and seem to have no observable educational value to the children. In other words, there is no reason to have a classroom animal except for regular educational purposes. One solution is a "borrowed pet" who visits at intervals from the home of one of the children or from a wildlife facility or park. The teacher may want to have the children generate questions about how the animal moves, gets its food, and protects itself. S. Simon's *Baby Animals* provides children with a wonderful introduction to small and large babies. It includes terms for baby animals and other interesting facts, along with full-color photos.

◇ Reflecting

Ask children to organize their experiences. Provide them with space so they can display their artwork, journals, and charts. Also exhibit their predictions, observations, and conclusions about various questions. Have children exhibit examples of the foods eaten by the animals they have studied. Provide poster paper for children to record differences in animal food preferences. If it is possible to display nesting materials, find an attractive space and add pictures and information on the type of bird, where it lives, and how it nurtures its young. Preserved insects may be added to the display.

Have a party! Invite family members and the school community to view the results of your active science experiences with small animals. (See Tear-Out Sheet 8-2.) Surprise them with questions about small animals, such as, "Is a spider an insect?" Maybe the children will be the experts. You might also provide suggestions to parents for supporting the science curriculum as described in the preceding section.

◇ Extending and Expanding to the Early Primary Grades

Although it is always difficult to classify concept development on the basis of age, children in the primary grades can usually do the following:

◆ Make better predictions about small animals and ask better and more specific questions.

- Conduct more sophisticated experiments. For example, they can investigate the housing, food, and habits of earthworms by raising them in the classroom. Further, they can investigate how the earthworms affect the soil over time. Other studies might include raising caterpillars and hatching chickens.

- Use field guides to identify birds and wildlife.
- Check out the websites of science museums and other resources for children. Teachers should make sure that the sites are suitable.
- Build a larger vocabulary of scientific terms pertaining to small animals and draw and label simple insects. These terms may be recorded in their science journals and posted next to experiments for easy reference.
- Investigate ponds. Many fascinating plants and animals live in, on, and around ponds. Visit a pond at different times of year to observe the changes.
- Classify and reclassify animals into more discrete groups. For example, instead of birds, insects, and animals that swim, children can form other categories such as where animals live, zoo animals, and endangered animals.
- Understand habitats. (See Chapter 7.)
- Learn about animals that hibernate. L. D. Brimmer's book *Animals That Hibernate* has excellent photographs.

[Date]

Dear Parents,

I hope you enjoyed our class party based on the theme **"Living Things Grow and Change."** I also hope that you enjoyed participating in the science activities and projects at home with your child. We are continuing with our theme about how living things grow and change by studying **"Insects and Small Animals."**

I will be sending home some information about guess what: bugs! Yes, I know that some bugs are real pests, but you and your child can learn by observing them. I will provide you with materials so that you don't have to handle the bugs directly. You will also need a picture insect guide from the library as well as your child's science journal. I do have some extra guides (in English or Spanish) that you can borrow.

Here are some steps to follow when studying the insects with your child:

1. Search your home and neighborhood for bugs. They may be anywhere. Have your child record in the journal where you find them.

2. Try to identify the types of bugs you find using the guides. Record these in the journal using diagrams or pictures.

3. Identify the questions that you and your child have about the insects.

4. Write down possible answers to the questions in your child's journal or just draw pictures of what you see to be analyzed later.

Thank you for helping with our new science unit about insects and bugs. I can always use volunteers in the classroom so please let me know if you are interested.

Sincerely,

Science Exhibit and Class Party

[Date]

Dear Parents,

We are about to conclude our science unit on **"Living Things Grow and Change: Insects and Small Animals."** I hope you have enjoyed the activities that you and your child completed at home. I know that your children have enjoyed all of the activities. Your participation and support helped the children learn many new science concepts. I really appreciate your support with this project.

Now it is time to have a party to recognize all of the hard work that you and your children have done in the area of science. You and your family are invited to our party at___________ on _______. We will have child care for younger children. Please let me know if you need help with transportation.

The class party will also include exhibits of all types, including charts, drawings, and experiments done at school (our famous mealworms!). Also on display will be the children's artwork, posters of science words the children have learned, our work on identifying birds and nests, and observations of our "borrowed" pets. We will have an exhibit on foods that birds eat and we will of course have plenty of food and drinks for you and your children.

Thank you for helping your children learn about insects and bugs and I look forward to seeing you at the class party.

Sincerely,

Group Observation: Science Terms
(Insects and small animals)

Date: ____________________________

Center/Outside: __________________

Children's Names: ______________________

		Accuracy	
Science Terms Used	**Not at All**	**Some**	**Accurate**
1. thorax	________	________	________
2. antennae	________	________	________
3. chrysalis	________	________	________
4. larvae	________	________	________
5. habitat	________	________	________
6. life cycle	________	________	________
7. camouflage	________	________	________
8. hibernate	________	________	________
9. migrate	________	________	________
10. metamorphosis	________	________	________

Comments:

Individual Evaluations: Assessing Children's Science Skills
Living Things Grow and Change: Insects and Small Animals

Date: ______________

Name: ______________

DOB/Age: ____________

	Always	Sometimes	Never
Observes insects and small animals outside and in the classroom	______	______	______
Compares insects and small animals	______	______	______
Identifies similarities and differences	______	______	______
Learns scientific terminology for insect and animal parts and functions	______	______	______
Observes and records the life cycle of an insect and small animal	______	______	______
Identifies environments where insects and small animals thrive	______	______	______
Identifies various shelters insects and small animals make for their young	______	______	______
Identifies foods specific to different animals	______	______	______
Identifies stories that depict animals accurately	______	______	______
Classifies insects and small animals	______	______	______
Asks appropriate questions	______	______	______
Reaches appropriate conclusions	______	______	______
Constructs simple experiments (with or without help)	______	______	______
Uses field guides and reference books	______	______	______
Uses science journal regularly	______	______	______
Draws and labels the parts of insects and small animals	______	______	______
Exhibits appropriate respect for insects and small animals	______	______	______

Notes/Comments:

9

How Toys Work

FOR THE TEACHER

◇ What You'll Need to Know

Children's natural curiosity and need to know drive them to learn more about the physical world in which they live. By studying things in their physical world and finding out how things work, children's scientific skills of observing, questioning, planning and conducting investigations, and reaching conclusions continue to develop.

You might begin with children's toys, many of which use a source of energy. The source may be the children who move them up and down on inclined planes or ramps or up and over bumps and boards. Spring toys use the energy stored in the spring, which when released, powers the toy. Electrical energy is stored in batteries, and the wind powers other toys.

Some toys make sounds. These sounds emanate from the vibrations of a part of the toy. Children can explore and feel these vibrations as they play with drums and other musical instruments.

Toys that are broken or no longer in working order can be used to find out how things work as well. After making certain that all toys are safety-proofed and that adult supervision is constant, provide children with screwdrivers, pliers, and other tools so they can take apart broken toys to find out how they work and perhaps even fix them (with assistance).

◇ Key Concepts

- Moving toys have a source of energy.
- This source may be the children themselves or may stem from springs or electricity.
- Wind or air is the source of energy for other toys.
- Vibrations cause some toys to make sounds.

◇ Goals and Objectives

The children will be able to

observe how toys move;

experiment by moving human-powered toys on inclined planes, ramps, and rough surfaces;

compare sources of energy-human, spring, electrical, or wind;

explore toys that make sounds through vibrations.

◆ Standards Alignment

Key Concepts and Goals and Objectives are based on the *National Science Education Standards* content standards, grades K-4. Standard B: Physical Science (properties of objects and materials, position and motion of objects, light, heat, electricity, and magnetism).

Benchmarks for Science Literacy (The Physical Setting, Common Themes)

◇ What You'll Need

Consult a variety of books on how to teach science concepts through toys. Some of these books can also serve as resources for older children:

1. *Teaching Physics with Toys: Activities for Grades K–9* (1995) by Beverly A. P. Taylor, J. Poth, and D. J. Portman (McGraw-Hill). This resource assists teachers in conveying physics concepts through simple activities that use popular children's toys.
2. *Teaching Chemistry with Toys: Activities for Grades K–9* (2005) by Mickey Sarquis (Learning Triangle Press). This book covers the basics of chemistry and more advanced concepts of chemistry using current toys that children will love working with.
3. *Science in Seconds with Toys: Over 100 Experiments You Can Do in Ten Minutes or Less* (1998) by J. Potter (Wiley). This is also a good resource book to lend to parents. It covers every aspect of the science of toys and the games kids play.
4. *Color and Noise! Let's Play with Toys! Experiments in the Play Room* (2001) by J. Labb (Houghton Mifflin). What are toys made of? What makes them go? These are only a couple of the questions explored in this easy-to-use book.
5. *Mobiles: Building and Experimenting with Balancing Toys* (1993) by B. Zubrowski, Boston Children's Museum Activities Series (Morrow). This book suggests hands-on experiments and constructions for older children. It's a great resource for teachers.
6. *Exploring Matter with Toys: Understanding the Senses* (1996) by Mickey Sarquis, (Learning Triangle Press). Based on the successful National Science Foundation-funded Teaching Science with TOYS project, this book meets the requirements of the National Science Education Standards. Useful for the chapter on the senses as well.
7. *Everything Kids' Magical Science Experiments Book* (2007) by Tom Robinson. This is a fantastic book with activities that your students will love.
8. *Exploring Energy with Toys* (1998) by Beverly A. P. Taylor (Learning Triangle Press). This books shows how to convert energy. Simple machines built with LEGO kits demonstrate the principals of levers, gears, and pulleys.
9. *Investigating Solids, Liquids, and Gases with Toys: States of Matter and Changes of State* (1998) by Jerry L. Sarquis (Learning Triangle Press). This book includes 24 toy-based science activities.

You'll need a variety of toys, including

- wooden, plastic, or other toy cars, vehicles, dolls, and tops that require human power to be set into motion;
- windup toys, such as cars, dolls, or other toys with springs;
- toys that work with batteries;
- wood planks for ramps;
- drums and other musical instruments;
- materials to make parachutes-cloth, string, a weight, and perhaps a washer;
- materials to make pinwheels-a sharpened pencil, white construction paper, a plastic drinking straw, scissors, a paper fastener, and markers;
- kites;
- rhythm sticks.

You'll also need an expert in electricity. This could be an older child or a teenager-someone who would work with the children with a flashlight bulb and bulb holder, some single-strand bell wire, and a battery.

Children's Books

The following books can be used to motivate children and integrate children's literature with science experiences:

Lionni, L. (1987). *Alexander and the Wind-Up Mouse.* New York: Knopf.

Moss, L. (1995). *Zin, Zin, Zin! A Violin.* New York: Simon & Schuster.

Pelletier, A., & Nash, S. (2007). *Toy Farmer.* New York Dutton Children's Books.

Pinkney, J. B. (1997). *Max Found Two Sticks.* New York: Aladdin Paperbacks.

Rey, M. (1977). *Curious George Flies a Kite.* New York: Houghton Mifflin.

Rey, H. A., & Rey, M. (2002). *Curious George Visits a Toy Store.* New York: Houghton Mifflin.

Royston, A. (1991). *What's Inside? Toys.* New York: DK Publishing.

Sadler, W. (2005). *Construction Toys.* Chicago: Heinemann.

Sadler, W. (2005). *Toy Cars.* Chicago: Heinemann.

Sadler, W. (2005). *Toys with Springs.* Chicago: Heinemann.

Websites

1. *The Science Toy Maker* website (http://www.sciencetoymaker.org) is a resource for teachers and parents, with project ideas for making science toys.
2. *Science Discovery* (http://www.teachpreschoolscience.com/ToolsandMachines.html) is a resource for preschool and kindergarten teachers. The site is a compilation of 75 core discovery experiences that are interesting and fun.
3. The *Terrific Science* website (http://www.terrificscience.org) includes a wealth of resources and ideas for teaching science.

The Home-School Connection 

Let parents know you are studying how toys work. Ask them to expand children's experiences by exploring how other things work at home. (See Tear-Out Sheet 9-1 at the end of this chapter.) This may also be a good opportunity for parents to discuss "wants and needs" with children. Together they can explore which toys are likely to last and prove to be worth the money.

◇ Documenting and Assessing Children's Learning

Conduct structured interviews with individual or small groups of children. Give them some toys that move by human, wind, electric, or spring power. Include a drum and rattle. Ask the children which toys they like and start a discussion about how toys work. Let the children play with the toys and talk about them.

Then ask the children to do the following:

- Identify one toy that is powered by electricity, one by children, and another by wind. Ask children why they think each toy works.
- Play the drum and feel the vibrations. Ask children if they know any other toys or instruments that make sounds through vibrations. Record their answers.

The teacher of the 4-year-olds, Mr. Green, began documenting an investigation on a chart this way: "The children began experimenting with ramps using floor boards in dramatic play. Joe found a wooden ramp and bucket of balls outside our classroom. These included ping-pong balls, golf balls, tennis balls, baseballs, and softballs. The children worked together to try out different ideas with the ramps and balls." Then he used his digital camera to record the various experiments conducted by the children. Children drew pictures and wrote stories about the ramps and balls. Eventually, all aspects of the investigation were charted and posted in the classroom.

Chapter 7 has a more general discussion of ways to document and assess the results of children's investigations. See Tear-Out Sheet 9-2 at the end of this chapter for an individual evaluation of children's concepts of toys, motion, and gravity.

FOR THE CHILDREN

Much of children's understanding of the energy sources that power their toys stems from their incidental experiences. Think ahead about the teaching strategies to use to foster children's exploration of how their toys work.

1. Sources of Energy

◆ *Children as a source of energy:* As children play with cars, trains, or trucks they push themselves, provide them with two or more wood or plywood planks. Put a wedge under one plank to create a slanted road. Buttress this roadway with another plank. As children roll their cars and trucks down the plank, ask them the following questions:

- How far did the truck/car roll?
- Could you make it go farther? How?

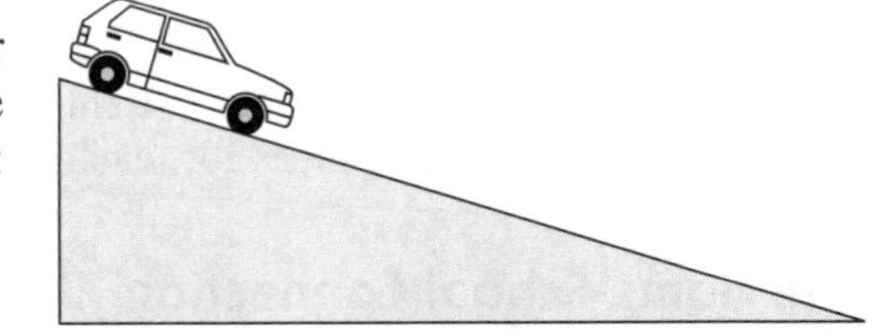

Then have the children roll one car along the floor and another car down the ramp, and ask them the following questions:

- Which car rolls the farthest?
- Which car goes the fastest?

Next, tie a piece of rope or heavy string to a few vehicles, and ask this question:

- Can you make the car roll up the plank?

Now have children pull them up the ramp.

While you're playing outside, take a few vehicles, planks, and rope with you. Have children try to roll their cars in sand, over the sidewalk, and through the grass. Ask them to compare and contrast their experiences.

- On what surface is it easier to pull or roll the cars?
- What causes the difference?

Have children record their experiences in sketches and drawings.

◆ *Springs as a source of energy:* Find a few vehicles that use springs as the source of energy. After children wind up the spring, ask them to repeat some of their experiences with cars they moved themselves.

- Have children wind up the vehicles and race them while other children push. Ask children to measure which goes the farthest and the fastest.
- Drive the spring-driven cars outside, on concrete, sand, gravel, and other surfaces. Compare the spring-driven cars with those pushed by hand over the same surfaces.
- Take apart one of the cars so that children can see the spring. Have them watch the spring as they wind it and as it releases. Relate the spring to the motion of the car.
- Find other toys that use springs as sources of energy. You might find a windup doll or animal.
- Read L. Lionni's *Alexander and the Wind-Up Mouse*, a magical story about a windup mouse.

◆ *Electricity as a source of energy:* With an expert in electricity to help you, ask children to experiment with lighting a flashlight bulb. Connecting wires to a bulb holder and to the two points on a battery will light a flashlight bulb. Have several sets so that a number of children can experiment at the same time.

- Find toy cars, other vehicles, dolls, or other battery-powered toys. Take these apart to show children the batteries. Children ages 4 and 5 might be able to learn the + and - signs and to match these signs on the battery to those in the toy.
- Provide children with a few small flashlights. Show them how the batteries are inserted into the light to turn the bulb on.

◆ *Wind as a source of energy:* Moving air is wind. Wind is caused by warm air rising over cool air. Children cannot see the wind, but they know when it's a windy day. Wind makes things move. Young children have some difficulty understanding how the wind produces electrical power and other things, but they can experiment with catching the wind and observe the effects.

Children take turns flying kites on a windy day.

- Make a couple of parachutes to take outside: Take a piece of lightweight cloth or a piece of plastic bag about 12 inches square. Punch a small hole in the center of the cloth and in each corner. Tie a string to each corner of the cloth. Tie all four strings together near their free ends. Fasten a small weight such as a metal washer to the free ends of the string. Once outside, show the children how to fold the parachute by holding the center and rolling it toward the strings. Then wrap the strings around the cloth. Have the children toss the parachutes into the air as high as they can. After children find out what they can do with the parachutes, try these experiments:

  Who can toss the parachute the highest?

  How far do parachutes fly?

  How many times can they jump on both feet before a parachute hits the ground?

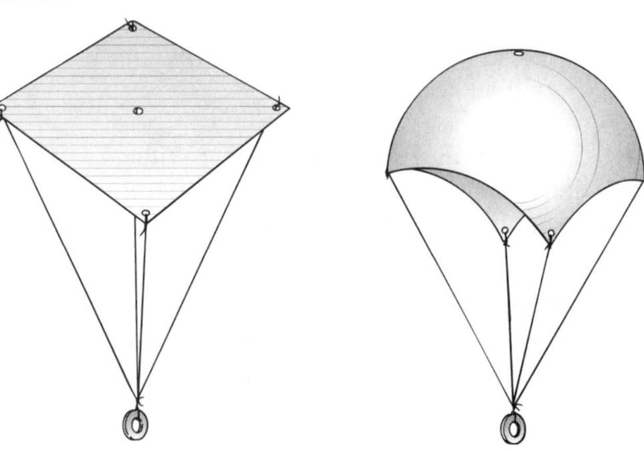

- Ask children to look at the parachutes and describe how one looks when it is first tossed in the air, when it fills with air, and when it falls to the ground. Back in the classroom, ask children to draw themselves tossing parachutes in the air (Seefeldt & Galper, 2000).
- Make or obtain pinwheels. For directions on making a pinwheels and a pinwheel pattern, go to http://janbrett.com. Young children haven't the muscle coordination to do this but will enjoy playing with the pinwheels outside in the wind.
- Obtain a few kites and take turns flying these on a windy day. As you do so, talk about how the wind feels as it tugs the kite high in the sky.
- Read M. Rey's *Curious George Flies a Kite.*

2. Vibrations

A group of children's toys make sounds. The sounds are usually caused by vibration, which in turn causes the air to vibrate. Observe toys that make sounds and try to discover the source of the sounds:

◆ Make rattles. Give each child two clear plastic glasses. Have them place a couple of sound makers (beads, seeds, pebbles) in the bottom of one glass. Help children tape the two glasses together to form a rattle. As children watch the objects make sounds as they move in the rattle, ask children to speculate on what makes the sounds.

◆ Make a drum by fastening paper over the open end of a large empty coffee can or oatmeal container. The paper can be held in place using large rubber bands. Beat the drum. While doing so, ask children to lightly place their hands on the drumhead so that they can feel the vibrations. Sprinkle some seeds on top of the drum. As children beat the drum, the seeds will dance from the vibrations of the drumhead. Have them speculate on why the seeds move. You do not need to lecture or tell children what is happening; all you want is for them to observe and question.

- Invite an older student to play a string instrument such as guitar or violin for the children. Ask the musician to show children how to vibrate the strings of the instrument to make sounds and to let children feel the instrument-and hence the vibrations-as they listen to the sounds produced.
- Create sounds using palm pipes. Palm pipes are made from different lengths of half-inch PVC pipe. You can either purchase a set of palm pipes or make your own by cutting different lengths of pipe. The students hold the pipe in one hand and strike one of the open ends on the palm of the other hand, producing the sound. Have students listen and observe each other as they take turns vibrating their palm pipes. Ask the children, "How are the sounds similar and how are they different?"
- Read *Zin! Zin! Zin!: A Violin* by L. Moss. This book presents 10 musical instruments in a playful way.
- Read *Max Found Two Sticks* by J. B. Pinkney, the story of a boy who makes music by tapping two sticks on his thighs, the bottom of a bucket, and other things to make music. Have children model Max by making music with rhythm sticks. Remind them that sounds are made through vibrations and ask them to feel the vibrations of the sticks.

◇ Reflecting

Make a group mural titled *All About Toys*. Divide the mural into four parts and label the parts as follows:

children

springs

electricity

wind

Precut pictures of toys from magazines for children to paste under the appropriate category. Or ask children to draw their own pictures of different toys. Then initiate a discussion of the children's investigations of the position and motion of objects.

◇ Extending and Expanding to the Early Primary Grades

Although fascination with motion begins at an early age, older children begin to connect their early experiments with dropping things, knocking down towers, and sending cars down a long block with beginning theories about why and how things move.

- Older children can conduct investigations on the various ways in which inanimate objects move. Create a KWL chart to determine what children know about how things move. For example, they may know
  - that things move if they are pushed or dropped to get started;
  - that people can do the pushing or pulling;
  - that batteries can make toys go;
  - that an engine makes their family car go.

Children wonder how to construct a tall building with blocks that won't fall down.

- They may want to know more about the concepts of gravity and balance. For example:
    - Why does the ball go down the hill without an engine?
    - Why do different shaped objects move differently?
    - How does the kind of surface on which the ball is moving affect the movement?
    - How can we construct a wide or tall building with blocks that won't fall down?
- To find beginning answers to their questions, children can be provided with the raw materials to make observations, comparisons, and measurements. Older children can take their own pictures to document their investigations. They can further decide how to record their information and explain it to others. Discussions with adults and other children about what makes things move challenges their thinking about animate and inanimate objects and the differences between them.

Date:

Dear Parents:

At school we are studying how toys work. At home you can extend children's ideas of how things work by showing them how household items work. You might show your children how to change batteries in a flashlight or one of their toys and talk about the electricity that powers your tools.

To develop beginning concepts of gravity you might try the following activity:

- Gather a variety of items of different shapes, weights, and materials like a feather, a piece of paper, a small rock, or a small ball. Have your child drop the items one at a time from the kitchen table to the floor.

- Discuss how each item moves on the way down. For example, does it float down? Does it move quickly or slowly?

Ask your child to pick some other small objects to drop. How do these fall? Suggest that your child draw a picture of these experiences or dictate a story about them. They can be sent to school to make an exhibit.

Our investigation about how toys work may also provide you with an opportunity to evaluate various toys and decide if they will last or be worth the price.

Thank you so much for your interest in our science program. Children learn best when parents are involved.

Sincerely,

Individual Evaluation of Children's Concepts of Toys, Motion, and Gravity

Date: ______________

Name: ______________

DOB/Age: ____________

	Always	Sometimes	Never
Observes how toys move in space	______	______	______
Observes how other objects move in space	______	______	______
Asks appropriate questions about toys and the movement of inanimate objects in space	______	______	______
Initiates investigations into how things move	______	______	______
Gathers appropriate evidence	______	______	______
Experiments with human-powered toys on inclined planes, ramps, and rough surfaces	______	______	______
Makes comparisons between the movement of various objects	______	______	______
Makes comparisons about movement using different sources of energy	______	______	______
Explores toys that make sounds through vibrations	______	______	______
Exhibits an understanding of correct terminology	______	______	______
Uses words such as *motion* and *gravity*	______	______	______
Experiments with balance using blocks	______	______	______
Documents conclusions appropriately	______	______	______
Explains conclusions to others	______	______	______

10

The Earth: Water, Weather, and Space

FOR THE TEACHER

◇ What You'll Need to Know

Water and the water cycle are part of the physical setting that we call the earth. It takes years for children to acquire the knowledge they need to complete the picture—earth studies include temperature, states of matter, and chemical concentrations in addition to water. This experience concentrates on concepts of water, yet learning relevant to other earth topics such as the study of weather and space is integrated into the learning activities. Again, children will learn through observation, comparison, predictions, inferences, and the communication of conclusions. Gradually, through active experiences with water and its properties, children form the foundation on which later abstract concepts about the earth can be built.

Because all living things require water to survive, our earth could not exist without it. Water is a readily available material that young children enjoy, and as part of their study, they should be helped to learn that it is also a precious resource to be conserved. Children make discoveries through unorganized water play such as pouring, splashing, and running through the sprinkler in the summer. As with sand, clay, and blocks, children can use water without being constrained by the one right way to use it. Yet when they are actively investigating a problem, they will engage in more sustained learning activity.

Trundle and Troland (2005) point out that the moon's cycle of phases is one of the most familiar natural phenomena, yet also one of the most misunderstood. Even elementary students who have a clear understanding of the shapes and relative positions of the sun, earth, and moon are unable to use this information to explain moon phases. Further compounding the problem, teachers often use children's literature, which misrepresents the moon in illustrations and text. Older children may create a Moon Calendar to record their observations and then compare them to the data in children's stories. For younger children, it may be best to use nonfiction books that treat lunar phases accurately.

◇ Key Concepts

- Water can be liquid or solid and move back and forth between the two.
- If water is turned into ice and the ice is allowed to melt, the amount of water is the same as before freezing.
- Water left in an open container disappears, but water in a closed container does not.
- Water has weight.
- Water's weight and upthrust help things float.
- Water is a solvent for many materials.
- Water moves into other materials and may change their composition.
- Atmosphere surrounds the earth.
- There are more stars in the sky than anyone can easily count, but they are not scattered evenly and are not the same brightness or color.
- The sun is seen in daytime.
- The moon is seen at night and sometimes during the day.
- The sun and moon appear to move slowly across the sky.

◆ Standards Alignment

Key Concepts and Goals and Objectives are based on the *National Science Education Standards* content standards, grades K–4. Standard B:Physical Science (properties of objects and materials, position and motion of objects, light, heat, electricity, and magnetism); Standard D: Earth and Space Science (properties of earth materials, objects in the sky, changes in earth and sky).

Benchmarks for Science Literacy (The Physical Setting, The Living Environment).

- The moon looks a little different every day for 4 weeks.
- Weather changes from day to day and over the seasons.
- Weather can be described by measurable quantities such as temperature and rainfall.

◇ Goals and Objectives

The children will be able to

observe and describe water in its various forms;

feel, smell, and taste water;

experience the weight of water;

compare which objects will float and which will sink;

observe air moving and air evaporating water;

observe how temperature change transforms water from liquid to solid and the reverse process;

discover which of a variety of materials dissolve in water;

identify materials that absorb water;

discuss the importance of water to life on earth;

measure water using various containers;

discuss the importance of conserving water;

observe that clouds differ from time to time;

experience and describe different types of weather;

describe how the weather affects them;

describe the day sky and the night sky and note the positions of the sun and the moon.

◇ What You'll Need

1. *Liquid Explorations* (1990) by L. Alger (Lawrence Hall of Science). This book describes liquid properties and is organized for classroom use.
2. *Exploring Water with Young Children* (2005)by I. Chalufour and K. Worth (Redleaf Press). It is also possible to purchase a trainer's video and guide. Part of the Discovering Nature series, this set focuses children's explorations to

deepen their understanding of water and its properties. Emphasis is on inquiry skills.

3. *Will It Float or Sink* (2006) by M. Stewart (Children's PR). This is a great introductory tool to teach young readers the mystery of flotation. The author incorporates colorful photos and simple text.
4. *The Science of Water: Projects and Experiments with Water Science and Power* (2002) by S. Parker (Heinemann). This book teaches the fundamentals of physics through hands-on projects that use everyday materials.
5. *The Science of Air: Projects and Experiments with Air and Flight* (2005) by S. Parker (Heinemann). This book contains 12 experiments with step by step photo instructions. Topics include air movement, air pressure, and wind resistance.
6. *What Is the Moon Made of? And Other Questions Kids Have About Space* (2009) by D. H. Bowman (Capstone Press). If you or your students have questions such as "How hot is the Sun?" or "Why is the planet Mars red?" this book has the answers for you.
7. *The Moon and You* (2001) by E. C. Krupp and R. R. Krupp (Harper Collins). This book contains a wealth of information about the moon's physical nature, phases, orbit, and eclipses. The book includes moon legends and beliefs among various cultures.
8. *Constellations, Stars & Planets, and Weather* (2005) by National Geographic (part of the My First Pocket Guides Series). Concise texts filled with fun facts, these guides encourage independent investigations.
9. *The Usborne Book of Astronomy and Space* (2009) by L. Miles (Usborne). This book is filled with up-to-date facts about space.
10. *The Kids Book of Weather Forecasting* (2008) by M. Breen, K. Friestad, and M. Klineis (Williamson Books). This is a fun-filled book of things for your students to do in the field of weather forecasting or simply to be exposed to the different aspects of weather.

Children's Books

Ackerman, J. (1009). *Welcome Fall.* New York: Scholastic

Asch, F. (2000). *Water.* Boston: Houghton Mifflin Harcourt.

Berger, M., & Berger, G. (1995). *Water, Water Everywhere: A Book about the Water Cycle.* Nashville, TN: Hambleton-Hill.

Branley, F. M. (1997). *Down Comes the Rain* (Let's-Read-and-Find-Out Science, Stage 2). New York: Harper Trophy.

Clark, J., & Zaidi, N. (2003). *Baby Einstein: Water, Water, Everywhere.* New York: Hyperion Book Co.

Cousins, L. (2006). *Maisy's Wonderful Weather Book.* New York: Walker.

Earle, S. (2000). *Sea Critters.* Washington, DC: National Geographic Children's Books.

Hooper, M. (1998). *The Drop in My Drink.* New York: Viking Children's Books.

Jeunessa, G. (1990). *Water.* New York: Scholastic.

Landry, L. (2007). *Space Boy.* New York: Houghton Mifflin.

Levy, C. (2002). *Splash!: Poems of Our Watery World.* New York: Orchard Books.

Locker, T. (2002). *Water Dance.* Boston: Houghton Mifflin Harcourt.

Marzollo, J. (1996). *I Am Water* (Hello Reader, Level 1). New York: Scholastic.

Mckinney, B. S., & Maydak, M. S. (1998). *A Drop around the World.* Nevada City, CA: Dawn Publications.

Neye, E. (2002). *Water, Vol. 1.* New York: Grosset & Dunlap.

Relf, P. (1996). *The Magic School Bus Wet All Over: A Book About the Water Cycle.* New York: Scholastic.

Sherman, J. (2003). *Gusts and Gales: A Book About Wind.* Minneapolis, MN: Picture Window Books.

Sherman, J. (2004). *Splish! Splash!: A Book About Rain.* Minneapolis, MN: Picture Window Books.

Simon, S. (2003). *The Moon.* New York: Simon & Schuster.

Swanson, D. (2005). *The Wonder in Water.* Toronto, ON: Annick Press.

Sweeney, J. (1999). *Me and My Place in Space.* New York: Dragonfly Books.

Thomas, R. (2003). *Eye of the Storm: A Book About Hurricanes.* Minneapolis, MN: Picture Window Books.

Tomecek, S. (2001). *Sun.* Washington, DC: National Geographic.

Tomecek, S. (2003). *Stars.* Washington, DC: National Geographic.

Tomecek, S. (2005). *Moon.* Washington, DC: National Geographic.

Wilson, L. (1993). *What's Out There: A Book About Space.* New York: Grosset & Dunlap.

Websites

1. *Water Theme Activities and Crafts* (http://www.first-school.ws/theme/mini_theme/water-cycle-quality.htm) is a website for preschool and kindergarten teachers. The website includes water-themed educational activities, including printable crafts, activities, and coloring pages with easy-to-follow lesson-plan instructions and related resources.
2. *Water Science for Schools* (http://ga.water.usgs.gov/edu/index.html) is a website hosted by the U. S. Geological Service. The site includes information on many aspects of water, along with pictures, data, maps, and an interactive center.
3. The *National Aeronautics and Space Administration* (NASA) (http://www.nasa.gov/audience/foreducators/k-4/index.html) provides many excellent education resources for teachers and students. Discover lesson plans, videos, and other great education resources from NASA.

OTHER THINGS YOU'LL NEED

- A transparent water table is preferable, although most are lined. A list of water play enhancers is endless: for example, soup ladles, strainers, funnels, measuring cups and spoons, squeeze bottles, corks, sponges, medicine droppers, detergent, colanders, strainers, and food coloring. Children may bring some of the items from home.
- Large, transparent plastic containers
- Small, transparent plastic containers
- Different types of paper, including paper towels

Funnels have an important function in the water table.

- Small wading pool if outdoor space permits
- Access to a freezing compartment
- Tools to measure changes in the weather
- Moon calendar for recording moon data
- Styrofoam trays
- Hammer
- Large and small nails
- Large bucket
- Plastic smocks
- Magnifying glasses

The Home–School Connection 

There are many ways in which parents can assist their children in developing concepts about water, weather, and climate at home. Simply pointing out that moisture condenses on mirrors and windows at bath time and that ice melts if left out of the freezer helps reinforce school learning. Snow and rain provide opportunities to discuss the qualities of water, how it feels on the skin, and what happens if you roll in the snow or splash in a puddle. Parents can also point out how the weather affects people. Things to mention are clothing, school closing days, and types of activities. Going to the swimming pool or beach, parents can encourage children to tell them how it feels to float or displace water by walking through it. Have children observe, predict, test, and record their conclusions.

The teacher may want to send some ideas to parents as the unit on water, weather, and climate progresses. Because children will not be able to observe the night sky at school, families can help to study the moon and stars by taking children out and recording their observations. Parents may ask some questions to motivate children, such as "How does the moon look tonight?" and "Do you think it looks bigger than last week?"

Some simple activities to do with children at home include making bubbles of beautiful shapes and colors. Just mix dishwashing liquid with water in a tray. Use a drinking straw and try out different-sized bubbles. What happens when a dry object touches them? Why? Encourage children to draw pictures or write in their science journals. Finally, teachers may suggest to parents that although their children should be free to examine the properties of water, water must be conserved because it is essential to life on earth.

Simple experiments to test what floats can also be carried out easily at home, even in the bathtub. Try a wood block, a plastic cap, one piece of aluminum foil tightly squeezed together, and one spread like a boat. These experiments should reinforce the concept that things float when the weight is supported by more water. Bath time is also an excellent time to use different-sized containers, spoons, and cups. Children will soon learn that water and other liquids take the shape of whatever container they are in.

◇ Documenting and Assessing Children's Learning

Documenting and assessing children's concepts of water and the water cycle will be a continuous process done on an individual and group basis using

- teacher observations of children's interest level in the form of anecdotal records.
- individual and group discussions with children as they work by themselves or in groups.
- structured interviews with children about the concept under study.
- portfolios of children's work, which will include drawings, stories, charts, and photographs of the children at work.
- examination of children's science journals. Older children should be encouraged and motivated to write in them each day.
- children's self-evaluations of their interest level, their work, and dictated stories about water.

Tear-Out Sheets 10-1 and 10-2 at the end of this chapter can be used at different times in the school year to chart growth in children's knowledge of water, the water cycle, weather, and space and to help plan a curriculum to extend and expand upon children's current understanding.

KWL charts, T-charts, and displays will document what the children knew before the investigations and what they knew after. Older children can make a chart of the water cycle, beginning with "How Rain Forms." All children could be asked to find an interesting fact about water, weather, and space. These facts may be displayed as a mobile hanging in the classroom or illustrated on a bulletin board or mirror. Children at the Head Start Center made a large chart to document what they knew about water. The teacher put "WATER" in the middle. Then each child dictated a sentence describing what he or she knew about water. The sentences went out in all directions with the child's name added at the end. It looked like following figure.

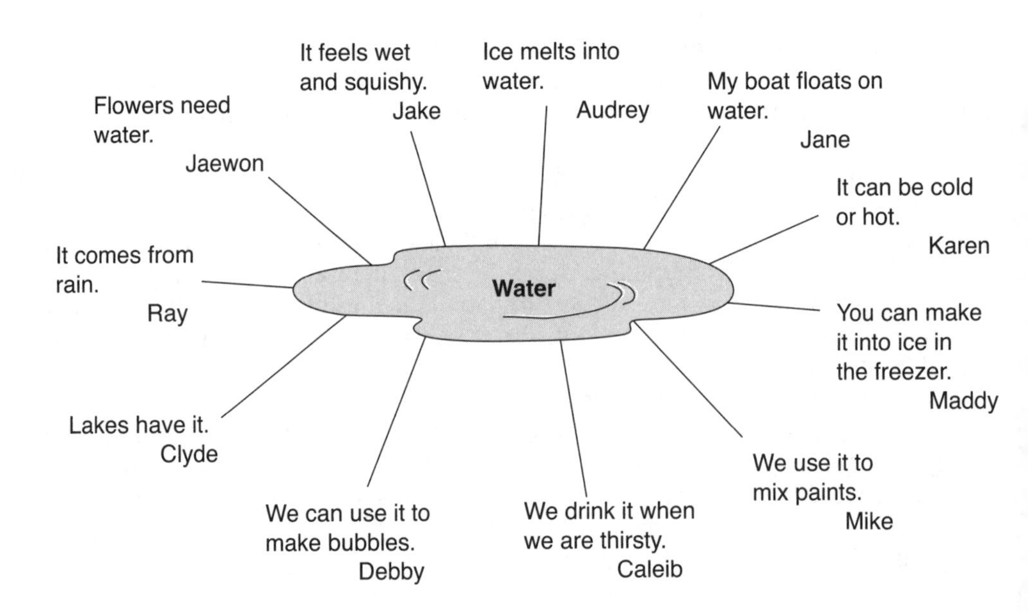

FOR THE CHILDREN

◇ Indoor and Outdoor Activities

The teacher should emphasize the theme of water, weather, and space throughout the classroom by using displays of children's books and reference books, posters, and photographs and by using opportunities to experiment with water both inside and outside. Water play can be introduced by filling the water table. Children will gravitate to a table or container that has been supplied with the many items described previously. To stimulate children's construction of knowledge, teachers may want to ask questions such as the following:

What does the water look like?

What color is the water?

What happens when you move your hands in the water?

Children should also drink clean water and be encouraged to discuss how it tastes. Children's answers will vary depending on their age and concept level.

- Have children make a rainmaker by punching holes in an empty plastic milk carton. Encourage them to make an interesting pattern of holes. Then have them fill the container with water and slowly raise it in the air to make rain. Children will learn about texture by feeling the water as it sprinkles down. Encourage them to articulate what they are observing.
- In many classrooms, children are familiar with recording the weather. Teachers can extend this experience by assisting children with collecting data and constructing graphs and charts that reveal weather patterns and seasonal changes. Then children will be able to answer such questions as, "How many sunny days did we have in November?"
- Display *A Drop of Water: A Book of Science and Wonder* by W. Wick (1997) and read sections of it to small groups of children. The wonderful photographs will set the stage for children to view water as rainbows, bubbles, drops, steam, ice, and frost.

Most impressive is a snowflake magnified to 60 times its actual size. Next, teachers may want to arrange the classroom so that children can construct their concepts about water.

- As an outdoor activity, arrange various-sized containers around a small pool. Ask children if water can be heavy. Suggest that groups of children try holding different-sized containers filled with water. Children can chart how heavy the water feels (relatively) in different containers. They can draw pictures and record their findings in their science journals. The same objective may be achieved by using a water table inside or having the children leave containers outside when it is raining and then try holding them.

- Have children compare which objects float and which sink in water. (Also suggested as a home activity.) Use dishpans of water and supply the assorted objects listed earlier. Children may participate by bringing some of the items from home. The teacher may want to ask such questions as, "Will the cork float?" Children draw their conclusions and record them. They can then try to see if the cork can stay on the bottom of the dishpan. Children can then classify objects as "Floaters" or "Sinkers." The teacher may want to assist children in charting their findings for display.

- Have children discover whether shape plays a part in whether an object is a "Floater" or a "Sinker." This can be with dishpans of water and aluminum foil in two shapes (tightly hammered and spread out). Children's findings may be added to the chart and journals.

- Using assorted wet and dry materials, help children to discover which will dissolve in water and which will not. Try salt, cornstarch, flour, salad oil, and sand. Suggest that children stir if they wish. Chart the findings under "Things that Dissolve in Water" and "Things that Do Not Dissolve in Water." Materials may be added as children become more sophisticated in their observations and conclusions.

- Depending on the location of the school, teachers can use resources to assist children in constructing the concept that changes in temperature transform water from liquid to solid and vice versa. Children can fill plastic bowls with water and put one bowl in the freezer and one in the room. Or, if the weather is below freezing, one container can be placed outside. Children should make predictions about the fate of the two bowls. Allow enough time and let the children compare the properties of the water.

- To compare the sun to the shade, try this sun and shade science activity. Take six bowls made of the same material. Place a cube of butter in each of the first two, a few ice cubes in the next two, and some old crayons in the last two. Then place a bowl with each item in the hot sun and the other three bowls in the shade.

- Have the children create a filter to clean polluted water. Punch some holes with a nail in the bottom of a cut-off milk carton. Spread some cotton inside on the carton bottom. Add some clean sand and put crushed charcoal on top of the sand. Then add another layer of sand. Place the filter on top of a small glass jar. Pour some clean tap water into your filter (this will pack the materials more tightly together). Prepare a jar of soil water: Put a handful of soil into a jar of water and mix. Let the water settle. Pour some soil water from the top of the jar into the filter. Watch the filtered water trickle into the small jar.

- Help children understand the concept of evaporation by having them soak a paper towel in water. Then squeeze the wet towel to remove excess water. Open the towel and lay it on a pie plate. Leave the towel and plate on your desk, and every so often, have the children feel the paper to see whether it is dry. Try putting one paper towel in a sunny location and another in the shade. Which paper towel will dry first?

- To prepare children for the concept that water moves into other materials on earth, ask them what will happen when they put water on different materials on their trays. Try tissues, paper towels, smooth paper, stones, and bits of fabric. Children can record their findings, which they later generalize to geology. They may also conclude that water will flow faster or slower and in different directions, depending on the material.

- To observe aquatic life in ponds and streams, have the children make an underwater viewer. They will need a teacher's help with the cutting but not with the observing that it makes possible. Using a plastic half-gallon milk or water container, carefully cut away the top, leaving the handle. Cut away the bottom of the container. Children may decorate it with water themes as they wish. Give children a piece of plastic wrap to stretch tightly over the bottom hole of the container, and hold it in place with a rubber band. To use the viewer, hold it by the handle and press it underwater so that the water comes up the sides but not into it. Peering down through the top, children will view slightly magnified creatures and plants.

- Read the short poem "Rain" by Robert Louis Stevenson. After the next rain, take magnifying glasses, science journals, and a marker outdoors to record the places where children find raindrops clinging. After returning to the classroom, have children generate a list of all the possible places that rain falls.

- Read *Down Comes the Rain* by F. Branley. Have children draw or paint a picture and dictate a story about how falling rain makes them feel and how their lives may change because of the rain. Jimmy may not be able to go out to play. Jane will have to put on boots and a raincoat.

◇ Reflecting

Ask children to organize their experiences by providing them with a large area in the classroom where they can display their journals, charts, and explanations of experiments with water. The water table might serve as a focal point, with objects available to illustrate principles of water. Sections of the room may be reserved for observations of the night sky (done with parents) as well as data collected and recorded about weather. Families will enjoy finding out how many rainy days there were in the spring.

Have a party for family members to view and experiment with active science experiences involving water. Parents and siblings will enjoy comparing which objects float and which sink, for example. Children may create charts with simple directions for their guests to follow. Or they may pose questions such as "Which of these materials will dissolve in water?" They may provide sheets for family members to record their conclusions. Older children may exhibit a mural, drawings, or posters about our natural environment and the importance of conserving it. Serve food to guests at your science party with plenty of water to drink. (See Tear-Out Sheet 10-3.)

◇ Extending and Expanding to the Early Primary Grades

Children in the early primary grades can do the following activities:

- Understand a water cycle simulation. Materials needed include ice cubes, two metal trays (one for the ice cubes and one to catch the rain), and a heat source. Because the kettle or other source of water vapor will be very hot, the teacher will need to assist small groups of children with this project. Ask children what they think will happen when the water vapor hits the tray with ice cubes in it. Record their answers. Then hold the source of water vapor under or beside the tray with the ice cubes. Water droplets will be created on the outside of the tray. These droplets will fall down into the pan like rain. Have children write their conclusions in their science journals.

A model of the solar system helps older children to visualize it.

- Read *The Magic School Bus Wet All Over: A Book About the Water Cycle* by P. Relf and *Water, Water, Everywhere: A Book About the Water Cycle* by M. and G. Berger. Discuss the books with the children and have them draw and label their impressions of how the water cycle works. Discuss the importance of water for living things.

- Teach students to love the earth and conserve its resources. Make frequent short trips outdoors into the natural environment. Model caring and respect for the natural environment. Explain that when too many waste materials are emptied into lakes, streams, and rivers, living things die or have less water. The children will be interested that no way has ever been discovered to make new water. Have them initiate an investigation into the ways in which water is cleaned in your area. A simple experiment in hand washing will show children how much water is wasted when they let it run. Just use a container to catch the extra water.

- Check out the websites of science museums and other resources for children. Teachers should make sure that the sites are suitable.

- Build a larger working vocabulary of scientific terms. These may be recorded in students' science journals and posted near experiments for easy reference.

- Create a bulletin board on the investigation of water.

Individual Evaluation: Assessing Children's Science Skills
The Earth: Water, Weather, and Space

Name: ______________

Date: ______________

DOB/Age: ______________

	Always	Sometimes	Never
Observes water in all its forms	_____	_____	_____
Feels smells and tastes water	_____	_____	_____
Compares the weight of water in different containers	_____	_____	_____
Compares objects that float and sink	_____	_____	_____
Observes under what conditions evaporation occurs	_____	_____	_____
Observes and describes how changes in temperature transform water from liquid to solid and the reverse	_____	_____	_____
Discovers what materials dissolve in water	_____	_____	_____
Determines which materials absorb water	_____	_____	_____
Discusses the importance of water to life on earth	_____	_____	
Discovers how much water is wasted in the classroom	_____	_____	_____
Visits ponds and rivers and observes water life	_____	_____	_____
Reads books about water, weather, and space	_____	_____	_____
Consults reference books about water, weather, and space	_____	_____	_____
Writes stories about water, weather, and space	_____	_____	_____
Helps in cooperative projects about water, including the bulletin board	_____	_____	_____
Asks appropriate questions	_____	_____	_____
Observes the day and night sky	_____	_____	_____
Records observations about the sun, stars, and moon	_____	_____	_____
Discusses weather changes	_____	_____	_____
Measures and records the weather	_____	_____	_____

Additional/Comments:

Group Observation: Science Terms
(These Terms Relate to Water and the Water Cycle.)

Date:__________________

Center/Area:__________________

Children's Names:__________________

	Accuracy		
Science Terms Used	**Not at All**	**Some**	**Accurate**
1. liquid	______	______	______
2. dissolve	______	______	______
3. evaporate	______	______	______
4. absorb	______	______	______
5. conserve	______	______	______
6. rain	______	______	______
7. water cycle	______	______	______
8. freeze	______	______	______
9. float	______	______	______
10. sink	______	______	______

Comments:

Science Exhibit and Class Party
Everyone Welcome!
Child Care and Refreshments

[Date]

Dear Parents,

We are about to conclude our investigation of **"The Earth: Water."** As you know, we have also included some observations of weather and space. During this period, we have focused on the importance of water to living things and of conserving water. We have conducted many experiments, both inside and outside, to learn about the properties of water, where it goes, and what it can do. I hope you have enjoyed the activities that you and your child completed at home. I know that your children have enjoyed all of the activities. Your participation and support helped the children learn many new science concepts. I really appreciate your support with this project.

Now it is time to have a party to recognize all of the fun and hard work that we have had learning about water. You and your family are invited to our party at _____ on ____. We will have child care for younger children. Please let me know if you need help with transportation.

The class party will also include exhibits of all types, including charts, drawings, and experiments done at school (our famous mealworms!). Also on display will be the children's artwork and posters of science words that the children have learned. We will have several fun science activities and will of course have plenty of food and drinks for you and your children.

Thank you for helping your children learn about water. I look forward to seeing you at the class party.

Sincerely,

11

The Earth: Rocks, Minerals, and Fossils

FOR THE TEACHER

◇ What You'll Need to Know

Through their explorations, young children are aware that the earth's surface is composed of rocks and various types of soil. By carefully observing and describing the properties of many rocks, children will begin to see that some rocks are made of a single substance, but most are made of several substances. The substances can be identified as minerals; however, the origin of rocks and minerals has little meaning to young children and should be left to the later elementary years.

School yards, backyards, playgrounds, vacant lots, and parks are good study sites to observe a variety of materials that make up the earth. As children collect rocks, they will become aware that soil varies from place to place in color, texture, and reaction to water. It is a good idea to revisit sites often so that children gain an understanding that the earth's surface is constantly changing. They can also simulate some changes, such as erosion in a small tray of soil in or outside the classroom.

Earth science is not about memorizing words and facts. It is about investigating familiar objects to build a foundation for future learning. Learning about rocks and minerals allows children many opportunities to compare, classify, predict, test hypotheses, and communicate their findings to others.

◇ Key Concepts

- Chunks of rocks come in many sizes and shapes, from boulders to smaller than a grain of sand.
- There are different types of rocks.
- Rocks are nonliving things.
- Rocks are formed in different ways.
- Sand is made up of tiny pieces of rocks.
- Rocks are composed of minerals, but the amounts of mineral vary from rock to rock.
- Rocks change by wearing away.
- Plants and animals left prints in rocks a long time ago.
- When rocks wear away and are combined with other materials, they produce soil.
- Minerals form crystals.
- The properties of rocks determine how they are used.

◆ Standards Alignment

Key Concepts and Goals and Objectives are based on the *National Science Education Standards* content standards, grades K–4. Standard D: Earth and Space Science (properties of earth materials, objects in the sky, changes in earth and sky).

Benchmarks for Science Literacy (The Physical Setting)

◇ Goals and Objectives

The children will be able to:

observe and describe various types of rocks;

describe rocks under various conditions;

test the hardness of rocks;

classify rocks by various attributes (size, shape, alike/different, color, smooth/jagged, hard/soft);

observe the process of making soil;

observe real fossils;

make fossil prints;

observe crystals;

participate in making crystals;

create art using pebbles/rocks, sand, and chalk;

draw, label, and record their observations about rocks and minerals;

learn new scientific terminology related to rocks and minerals.

◇ What You'll Need

Unless you are an avid rock hound, you should consult some reference books and have them available for you and the children as you plan your active experiences with rocks and minerals. Most of the following books for adults and older children have been recommended by the NSTA or may be found in *Science Books and Films*, published by the AAAS:

1. *Problems as Possibilities: Problem-Based Learning* (2002) by L. Torp and S. Sage (Association for Curriculum and Development). This book offers suggestions for posing problems. Each phase of problem solving is mapped out so that teachers know what they should do and what to expect of children.
2. *Explore Your World: Rocks and Minerals* (1999) by Discovery Channel (Random House). Incorporates the Discovery Channel's acclaimed visuals, with more than 300 full-color photographs and visuals of rocks and minerals.
3. *The Best Book of Fossils, Rocks, and Minerals* (2002) by C. Pellant (Houghton Mifflin). This book covers the topic of fossil rocks.
4. *My First Pocket Guide: Rocks & Minerals* (2005) (National Geographic). This guide is one of a number of National Geographic field guides created especially for grades 1 to 5.
5. *Kingfisher Young Knowledge: Rocks & Fossils* (2003) by C. Pellant (Houghton Mifflin). Another accessible book.
6. *Smithsonian Handbooks: Rocks & Minerals* (2002) by C. Pellant (Dorling Kindersley). This guide explains what rocks and minerals are and how they are classified.
7. *The Complete Illustrated Guide to Minerals, Rocks, and Fossils of the World* (2010) by John Farndon (Lorenz Books). This is a comprehensive reference to more than 700 minerals, rocks, plants, and animals.
8. *Experiments with Rocks and Minerals* (2002) by T. Salvatore (Bt Bound). This is a science experiment book for children of all ages that explains how minerals are different from rocks and why they are important.

9. *Rocks and Minerals: A Gem of a Book* (2009) by S. Basher (Kingfisher). This book is an in-depth look at the ground beneath our feet. The topics are presented through charming and adorable illustrations paired with basic information told from a first-person perspective.
10. *Extreme Rocks and Minerals* (2007) by D. Green and S. Basher (HarperCollins). This book is full of information and stunning photographs of extreme rocks and minerals.

Children's Books

Aliki. (1990). *Fossils Tell of Long Ago.* New York: HarperCollins.

Bailey, J. (2006). *The Rock Factory: The Story about the Rock Cycle.* London: Science Works.

Baylor, B. (1987). *Everybody Needs a Rock.* New York: Aladdin Books.

Christian, P. (2000). *If You Find a Rock.* San Diego: Harcourt Brace.

Ewart, C. (2004). *Fossil.* New York: Walker.

Gans, R. (2009). *Let's Go Rock Collecting*. New York: Harper Collins.

Kittinger, J. S. (1997). *A Look at Rocks: From Coal to Kimberlite.* Danbury, CT: Franklin Watts.

Lionni, L. (1995). *On My Beach There Are Many Pebbles.* New York: Mulberry Books.

Marzollo, J. (1998). *I Am a Rock* (Hello Reader Level 1). New York: Cartwheel Books/Scholastic.

Parker, S. (1997). *Eyewitness Explorers: Rocks and Minerals.* New York: D. K. Publishing.

Pellant, C. (2007). *The Best Book of Fossils, Rocks, and Minerals.* St. Louis, MO: San Val.

Rook, D., Coenraads, R. R., & Busbey, A. B. (2000). *Rocks and Fossils.* New York: Time Life Education.

Stotsky, S. (1998). *Geology: The Active Earth (Ranger Rick's Naturescope).* Broomall, PA: Chelsea House.

Websites

1. *Rocks for Kids* (http://www.rocksforkids.com) is a website for adults and children. Here you will find information about rocks and minerals and suggestions of where to go to find more.
2. Kathi Mitchel's website (http://www.kathimitchell.com/rock.html) has a large section on rocks and minerals with links to other websites. The site is designed for teachers and students to find information and ideas. .
3. The *RockhoundKids* website (http://www.rockhoundkids.com) is an excellent resource for teachers. If you're a kid or a kid at heart and love rocks, minerals, and geology in general, you'll love this website!

◇ Other Things You'll Need

- Rocks of all types. You may need to purchase or ask for donations of rocks of various sizes in bulk from building suppliers or landscapers.
- Containers of various types for storing, observing, and classifying rocks. Egg cartons and muffin tins are good for small collections.
- Charts and posters of rock formations.

- Hammers.
- Safety goggles.
- Newspapers.
- Empty cans.
- Water.
- Glue.
- Scales.
- Magnifiers.
- Tweezers.
- Different types of salts.
- Sieve.
- Trowel.
- Clay.
- Food coloring.
- Examples of minerals and fossils.

The Home–School Connection

There are many ways in which parents can assist their children in the development of concepts about rocks and minerals. Simply observing and collecting different rocks on a walk to the store or during playtime in the back yard will begin the process. If families live near a beach or a stream, they can not only collect rocks of different sizes and types but also have their children experience how the rocks feel when they walk on them. Both dry and wet rocks may hurt the feet, but wet rocks are slippery. How do the rocks look when they are dry and when they are wet?

Children love rocks and minerals and tend to collect them in pants and jacket pockets without suggestions from teachers and family members. Family members can supply simple containers to begin rock and mineral collections. Then they can suggest that children classify them in different containers according to various attributes, which will vary with the rocks that children found. Children may label the containers to indicate the basis of their classification.

Teachers may want to suggest that family members read the book *If You Find a Rock* by P. Christian (Harcourt, 2000). This book gives family members and children a variety of purposes for active learning experiences that are fun for all, such as using stones for skipping across ponds, finding tiny creatures under them, and climbing larger rock formations. The book deals with emotions as well, so parents may want to suggest that their children pick a rock that reminds them of a person or place that they care about. After all, rocks provide stability to our environment.

Finally, teachers may want to send family members the instructions for various active experiments. In *Helping Your Child Learn Science,* published by the U.S. Department of Education, there is a suggestion for at-home activities involving crystals. Children may use magnifying glasses to look for crystals and inspect table salt, Epsom salt, a honey jar, and the walls of the freezer (if it is not the frost-free kind). Children should draw pictures of what they see in their science journals. Parents can ask the children if all of the crystals look the same and, if not, how they are different. It is not necessary for family members to spend a lot of time locating rocks and minerals, however. Potential learning experiences present themselves every day.

◇ Documenting and Assessing Children's Learning

Assessing children's concepts of rocks and minerals is a continuous process done on an individual and group basis, using

- teacher observations in the form of anecdotal records and check sheets;
- individual and group discussions with children as they work by themselves or in groups;
- structured interviews with children about the concept under study;
- portfolios of children's work (photographs, drawings, charts, labels, artwork, terminology, and stories);
- examination of children's science journals;
- children's self-evaluations of their interest level, their work, and art and stories about rocks and minerals.

Child-made bulletin boards, mini-museums and labeled displays of all kinds, word walls, Venn diagrams, T-charts, graphs, and other forms of data collection all document children's new concepts of rocks and minerals. Chapter 6 gives a fuller explanation of the documentation and assessment process.

Tear-Out Sheets 11-2 and 11-3 at the end of this chapter can be used at different times in the school year to chart growth in children's knowledge of rocks and minerals.

FOR THE CHILDREN

◇ Indoor and Outdoor Activities

The teacher should emphasize the theme of rocks and minerals throughout the classroom using displays of children's books and reference books, posters, and photographs and by using opportunities to experiment with the properties of rocks and minerals both indoors and outdoors. The sand and water tables are excellent places to experiment with rocks and minerals. Simple concepts of erosion are formed as children blow on sand or pour various weights of water into the sand table.

Teachers suggest that children closely observe the rocks they have collected.

The following activities allow children to construct basic concepts of rocks and minerals. Make sure that child observations, predictions, classifications, and conclusions are recorded in science journals or charted for classroom reference. First, take children on a rock hunt. Provide them with containers to hold their favorite rocks. The play yard is a fine location for most rock hunts. More exotic rock forms will be found at a pebbly beach, by ponds and the banks of streams, and near large rock formations. You may need to provide examples of rocks that children will not find as part of your supplies from the landscaper. After the rocks are collected, you can implement various activities.

- Have children observe closely the rocks they have collected. Suggest that they use their senses—sight, touch, smell, and perhaps taste (if the rocks have been thoroughly washed)—to describe their observations. Provide children with magnifiers. After they have observed their rocks, have them share their descriptions. At this point, they will begin to build their vocabulary of descriptive words such as *rough, slippery, pointy, round,* and various colors. Make a chart with two columns. Ask children to tell you what they expected to see (column 1) and what they saw that they didn't expect (column 2).

- Have the children discover whether rocks look different from the inside. Cover a small rock with fabric and strike it hard with a hammer. Examine the inside; compare it with the outside appearance.

- Suggest that children sort rocks that are alike in some way into the same container. Let children make decisions regarding how they classify their rocks. Help them record the categories that they have devised and suggest others based on the descriptive words that children use.

- Although elementary-age children can use the test for hardness suggested in Roma Gans's book *Let's Go Rock Collecting,* teachers should use less dangerous materials for preschoolers and kindergartners. To determine which rocks are hard and which are soft, have children make scratch tests with their fingernails and with pennies. Rocks scratched by fingernails would be categorized as soft, rocks scratched by a penny would be in a middle category, and rocks that resist scratching by a penny would be the hardest rocks. Have children create and decorate containers to house the rocks and help them to label the containers.

- Read L. Lionni's *On My Beach There Are Many Pebbles.* Using pebbles previously collected or bought, have children discover how the appearance of pebbles changes if they are wet or dry. Have children view them in the water table. Then have the children remove some and check them the next day. Record children's explanations for the change in appearance. Show children how a drop of mineral oil or baby oil will transform washed and dried pebbles back to their original state. Children may use the colorful pebbles to make a centerpiece for their lunch table or to cover the soil in a flowerpot.

- Use rocks and pebbles and white glue to create a pebble sculpture. It can be representational or abstract—the form should be left to the children. Sometimes the shape or size of a rock or pebble will suggest a particular design. Make sure that children put their sculptures together before they use the glue so that (1) the design is pleasing to them and (2) it will sit solidly.

- Sand painting fits in well with activities centered on Native American culture. Have books with attractive photographs around the room. Using heavy paper, children will make a design with white glue. Suggest that they plan their design before they add the sand. Then sprinkle plain or different-colored sand on the paper and shake the excess onto newspaper. Older children may make more sophisticated designs by repeating the process until the paper is almost filled.

- Young children are familiar with chalk as an art medium, but they probably don't know that chalk is a rock. Children may draw with chalk on paper, old-fashioned "blackboards," and outside spaces where the rain will wash it away. Pencils also have components made from rocks and can be examined for their properties.

- Children are fascinated with crystals and enjoy making them with the help of the teacher. Dissolve salt crystals to make new ones. Dissolve 1 teaspoon of salt in 1 cup of water. Heat the mixture over a low flame to evaporate the water. The children will observe what is left. What shapes are these crystals?

- You may be lucky enough to find a fossil, but you will probably have to purchase or borrow one or take the children to a museum to see fossils. Explain that prints of animals, shells, and plants that lived on earth millions of years ago can be found pressed between layers of rocks. These rocks are called fossils. Using clay, children will enjoy making similar prints of their hands, shells, or leaves.

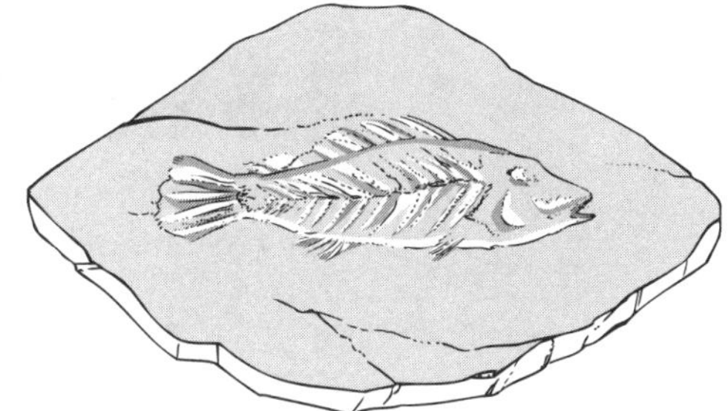

- Read *Fossil* by Claire Ewart. This story of how a fossil came to be is written as a moving and graceful poem. Have children choose a possible fossil from one of the guide books and write and illustrate a poem about it.

- Use plaster of Paris to create fossils. First, take a piece of plasticine modeling clay and create a flat, round shape about the size of a large cookie. Next, create a half-inch rim that will fit on the circumference of the clay cookie. Attach the rim to the clay cookie. Next, flatten out the top surface of the clay cookie into a smooth surface. Use a leaf or small shell to make an imprint in the clay. You could also use a pipe cleaner to make an imprint of a fossil in the clay. Carefully rub some Vaseline on the surface of the mold. Mix the plaster and carefully pour it onto the fossil mold. When the plaster has hardened, carefully remove the clay; you should be left with a fossil that can be painted. Bury them in sand or dirt and have an expedition to dig up fossils like paleontologists do.

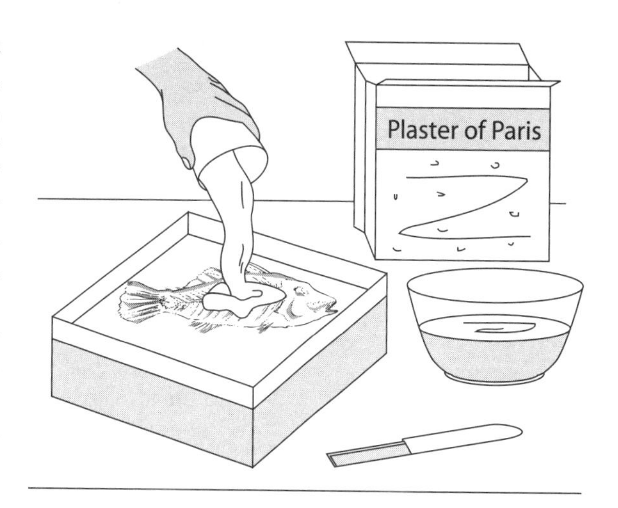

◇ Reflecting

Children can organize their experiences by using a large area of the classroom to display their journals, charts, and artwork. Mini-museums can be created to house beautiful rocks and minerals. The teacher may want to organize small meetings of "scientists" to discuss what they have been exploring and investigating.

Have a party for family members to view and experiment with active science experiences involving rocks and minerals. Everyone old and young alike will enjoy a variety of art activities and experiments that children have created as hands-on activities. Make sure that the children provide simple directions to follow. Give everyone a chance to

The teacher meets with a small group of "scientists" to discuss what they have been investigating.

admire the beautiful exhibits (labeled) and the charts of new scientific vocabulary words. (See Tear-Out Sheet 11-1 at the end of this chapter.)

◇ Extending and Expanding to the Early Grades

Children in the primary grades can do the following:

- Form some concepts about the three types of rock formations. By collecting rocks and figuring out how they were made, they can get a glimpse of how our planet was formed. Teachers should have reference books, posters, and examples of each of the following three categories of rock to use as they explain them and the children examine them:

  1. *Igneous* means "fire." This category of rock was once melted liquid deep inside our planet. Some kinds of igneous rock came flying up out of the earth in volcanic eruptions. Others cooled and hardened before they made it all the way to the surface but were exposed later by wind and rain. Granite, basalt, and obsidian are igneous rocks. Read *On the Spot: Volcanoes* by Angela Royston (Reader's Digest, 2000). This little informational book is full of visual effects and interesting details that children will love.
  2. *Sedimentary* rocks started out as silt, sand, or smashed-up seashells floating in water. Eventually, the tiny bits settled to the bottom of the water, where they built up in thick layers. The weight squeezed the grains together, and minerals in the water seeped into the spaces between the particles and bonded them into solid rock. Limestone and chalk are sedimentary rocks made from pulverized seashells. Have children draw with various types of chalk rocks on different papers to experiment with the effects. Ask them if they knew that the chalk they use every day was a rock. Have them use simple tools to smash sandstone. Ask them what they think sandstone is composed of.
  3. *Metamorphic* rock is igneous or sedimentary rock that has been pressure-cooked under the earth's surface to form a different and tougher rock. Provide

examples of limestone and marble. Marble is a rugged version of the "softer" limestone. Have children examine and chart the attributes of each of these rocks. Shale and slate may also be used to demonstrate the differences in metamorphic rocks.

- Read J. Marzollo's *I Am a Rock* (Scholastic, 2004) and enter the Rock Hall of Fame, where children will be provided with clues for identifying 12 rocks. (Labeled colored photos are at the end of Marzollo's book.)
- Learn more complex classification systems for rocks and minerals.
- Build a larger and more complex scientific vocabulary pertaining to rocks and minerals.
- Work with more complex tools such as knives and hammers to open rocks and minerals and observe their properties.
- Work with machines such as those that polish pebbles to observe the results of motion on the pebbles.
- Check out websites for information on rocks and minerals.
- Profit from a field experience to a science museum or the geology laboratory of a university. The teacher can preview the site in advance and have children generate the questions that they would like to have answered by the docent or student. When they return to the classroom, they can record what they have learned. Docents are increasingly often being trained to speak to children at their level, to answer their questions, and to introduce concepts that are not too advanced.

Science Exhibit and Class Party:
Everyone Come!
Child Care and Refreshments

[Date]

Dear Parents,

We are about to conclude our thematic unit on **"The Earth: Rocks and Minerals."** During this period, we have focused on the composition of the earth and the beauty and importance of various types of rocks and minerals. We have examined fossils and conducted many experiments, both indoors and outdoors, to investigate the properties of rocks, classify them, and see what happens to them under certain conditions. We hope that you have enjoyed the activities that you have done at home with your child to learn more about rocks and minerals, large and small. We also hope that you have enjoyed the books that we suggested. Please let us know if you have any book ideas for us. We are anxious to hear how your explorations with crystals turned out.

Now it is time to have a party to recognize all of the fun and hard work that we have had learning about rocks and minerals. Your whole family is invited, and transportation will be provided for anyone who needs it.

On _________ at _________ in the _________ at school, we will be having a science party. The event will include exhibits of all types, including your children's wonderful artwork with rocks and minerals. Rock drawings, posters listing the new science words we have learned, and experiments and art for you to do with your family will be featured.

There will be plenty of refreshments for us to enjoy. We thank you again for working with us and look forward to seeing you at the party.

Sincerely,

Group Observation: Science Terms
(These terms refer to rocks, minerals, and water.)

Date:_______________

Center/Area:_______________

Children's Names:_______________

	Accuracy		
Science Terms Used	**Not at All**	**Some**	**Accurate**
1. crystal	_____	_____	_____
2. fossil	_____	_____	_____
3. mineral	_____	_____	_____
4. soil	_____	_____	_____
5. erosion	_____	_____	_____
6. rough	_____	_____	_____
7. slippery	_____	_____	_____
8. chalk	_____	_____	_____
9. sand	_____	_____	_____
10. marble	_____	_____	_____

Comments:

Individual Evaluation: Assessing Children's Science Skills
The Earth: Water, Weather, and Space

Date: ________________

Name: ________________

DOB/Age: ________________

	Always	Sometimes	Never
Observes various types of rocks	_____	_____	_____
Describes rocks verbally	_____	_____	_____
Draws and describes rocks in science journal	_____	_____	_____
Tests the hardness of rocks	_____	_____	_____
Classifies rocks by various attributes	_____	_____	_____
Observes content of soil	_____	_____	_____
Observes fossils	_____	_____	_____
Makes fossil prints	_____	_____	_____
Observes crystals	_____	_____	_____
Participates in creating crystals	_____	_____	_____
Creates art using pebbles, sand, rocks, and/or chalk	_____	_____	_____
Reads books about rocks and minerals	_____	_____	_____
Writes or dictates stories about rocks and minerals	_____	_____	_____
Asks appropriate questions	_____	_____	_____

Notes/Comments:

12

The Human Body: The Senses

FOR THE TEACHER

◇ What You'll Need to Know

Children, like adults, use their five senses to find out about their surroundings and themselves. When educators say that hands-on activities are best in facilitating children's learning, what they often mean is that children should be encouraged to use all of their senses to take in information and utilize it to answer the questions posed by their environment. Different senses provide different information for the child. Younger children are capable of understanding only one or two attributes of something like an apple. When senses are coordinated, such as smelling and tasting, more information becomes available to form a concept of "apple." So older children should also be encouraged to understand how their senses work together. Through active experiences based on the senses, children may also gain an understanding of people who have impaired sensory abilities, such as hearing and visual impairments.

◇ Key Concepts

- There are five senses—hearing, smell, taste, touch, and vision.
- All of the senses can be used to find out about people, places, and things.
- People use their senses to find out about themselves.
- Used together, the senses can give us more information.
- It is necessary to practice using our senses so that they can help us learn more efficiently.
- Some people are not able to use one or more of their senses.

◇ Goals and Objectives

The children will be able to:

name the five senses;

tell how each sense helps us learn;

associate a part of the human body with each sense;

use one or more of the five senses to discover properties of objects in the environment;

compare objects using only one sense;

classify objects using only one sense;

describe how two or more senses work together;

appreciate the senses as a learning tool;

develop an understanding of persons with sensory impairments such as hearing or visual impairments.

◆ Standards Alignment

The Key Concepts and Goals and Objectives are based on the *National Science Education Standards* content standards, grades K–4. Standards D: Earth and Space Science (properties of earth materials, objects in the sky, changes in earth and sky).

Benchmarks for Science Literacy (The Physical Setting)

◇ What You'll Need

Consult some reference books and have them available for you and the children as you plan your active experiences with the senses. The following sources are recommended for adults and older children:

1. *The Science Book of the Senses* (1991) by N. Ardley (Harcourt Brace Jovanovich).
2. *Seeing, Smelling, and Hearing the World* (1997) by M. Pines (Howard Hughes Medical Institute).
3. *Kids Discover: The Five Senses* (1991, Vol. 1, No. 3) by S. Sands (Kids Discover). This magazine explores each sense separately with diagrams and photographs. It discusses the need to use the senses to experience life.
4. *Early Themes: 5 Senses (Grades K–1)* (1999) by T. West (Scholastic).
5. *The Five Senses* books by M. Rius, J. M. Parramon, and J. J. Puig (Barron's Educational Series). These books explore sight, hearing, smell, touch, and taste. They were first published in 1985 and most of them can be found in Spanish.
6. *Eyes and Ears* (2003) by S. Simon (HarperCollins). This book uses powerful photographs and illustrations to enhance written descriptions of how the eyes and ears work to sense the surrounding world, contribute to the sense of balance, and send messages to the brain.
7. *Understanding Your Senses* (2004) by R. Treays, E. Danes, and G. Doherty (Usborne Books). This book answers questions about sensational phenomena.
8. *The Sense of Smell* (2009) by E. Weis (Scholastic). This book describes the role of the sense of smell in human life, how people smell, smell and taste, the psychology of smelling, artificial scents, and conditions that affect this sense.
9. *The Sense of Taste* (2009) by E. Weis (Scholastic). This book provides surprising facts about our sense of taste. The book includes photos, timelines, and diagrams.
10. *The Sense of Touch* (2009) by E. Weis (Scholastic). This book provides facts about your sense of touch and how it is connected with all your other senses.

Children's Books

There are numerous children's books about the senses. A large list of references is presented in this section. Some of them have had multiple editions, but remain classics, such as the B. Martin, E. Carle, and H. Oxenbury books. The teacher must decide which books fit best with the children's investigations as they progress.

Aliki. (1990). *My Five Senses (Let's Read and Find Out).* New York: HarperTrophy.

Ballard, C. (1998). *How Do We Feel and Touch? (How Your Body Works).* Austin, TX: Raintree.

Ballard, C. (1998). *How Do We Taste and Smell? (How Your Body Works).* Austin, TX: Raintree.

Chancellor, D. (2007). *I Wonder Why Lemons Taste Sour and Other Questions About the Senses.* Boston: Kingfisher.

Ciboul, A. (2005). *The Five Senses.* Richmond Hill, Ontario: Firefly Books.

Cole, J. (2001). *Magic School Bus Explores the Senses.* New York: Scholastic.

Collins, A. (2006). *See, Hear, Smell, Taste, and Touch: Using Your Five Senses.* Washington, D.C.: National Geographic.

Cromwell, S. (1998). *How Do I Know It's Yucky? And Other Questions about the Senses (Body wise).* Des Plaines, IL: Rigby Interactive.

Hewitt, S. (1999). *The Five Senses (It's Science).* Danbury, CT: Children's Press.

Martin, B., Jr., & Carle, E. (1970). *Brown Bear, Brown Bear, What Do You See?* New York: Holt, Rinehart, and Winston.

Martin, B., Jr., & Carle, E. (1997). *Polar Bear, Polar Bear, What Do You Hear?* New York: Henry Holt.

McMillan, B. (1994). *Sense Suspense.* New York: Scholastic.

Nettleton, P. (2006). *Look, Listen, Taste, Touch, and Smell.* Minneapolis: Picture Window Books.

Oxenbury, H. (1985). *I Hear, I See, I Touch.* New York: Knopf.

Shannon, D. (2005). *David Smells!* New York: Scholastic.

Showers, P. (1990). *Ears Are for Hearing.* New York: Thomas Y. Crowell.

Showers, P. (1993). *The Listening Walk.* New York: HarperTrophy.

Stihler, C., & Rose, H. (2005). *Squishy, Squishy: A Book about My Five Senses.* Boston: Pauline Books and Media.

Sweeny, J. (2004). *Me and My Senses.* New York: Random House.

Walpole, B. (1997). *See for Yourself Series: Touch, Smell, and Taste, Seeing, and Hearing.* Austin, TX: Raintree.

Wood, N., & Rye, J. (1991). *Listen . . . What Do You Hear?* Mahwah, NJ: Troll Associates.

Wood, N., & Willey, L. (1991). *Touch . . . What Do You Feel?* Mahwah, NJ: Troll Associates.

Ziefert, H. (2002). *You Can't Taste a Pickle with Your Ear: A Book about Your 5 Senses.* Maplewood, NJ: Blue Apple Books.

Websites

1. The *Everything Preschool* website (http://www.everythingpreschool.com/themes/fivesenses/science.htm) has a whole section devoted to activities about the five senses.
2. The *Preschool Rainbow* website (http://www.preschoolrainbow.org/5senses.htm) includes a series of lessons and activities on the theme of the five senses.
3. *Littlegiraffes.com* (http://www.littlegiraffes.com/fivesenses.html) has a wealth of ideas for teaching about the five senses..

◇ Other Things You'll Need

Materials and equipment for assisting children in learning about and appreciating their senses are endless. They are listed in conjunction with each active learning experience described on the following pages.

The Home–School Connection

The home and its environs offer many opportunities for parents and children to explore the senses and share sensory experiences. Both inside and outside, there are things to see, smell, hear, taste, and touch. When parents encourage close observation through the senses, children learn the importance of their perceptual functions in concept formation.

Parents (and teachers) must formulate safety rules for sensory exploration. For example, not all things should be tasted or touched. Children should taste and touch only those items that parents have approved. Parents will probably be aware of their children's allergies, but teachers should certainly be informed. The senses also warn children about danger, but they may not be fully aware that a very hot burner in the kitchen will seriously burn them.

The following experiences require almost no preparation on the part of parents except perhaps time. Teachers may want to send these suggestions home in a handout or newsletter or discuss them at a meeting with parents:

- Parents can help children sharpen their listening skills. Read P. Showers's *The Listening Walk.* After discussing the things that parents and children may hear on a walk (especially if they are quiet), parents may venture out with their children to take a walk. Parents and older children should record everything they can hear. These sounds vary with the time of day and the location of their home. For example, city children will probably hear cars, airplanes, buses, and the chatter of people and businesses. Children in the country may hear the wind, rustling leaves, small animals, birds, and insects. As a follow-up activity (or for a parent meeting or party), teachers may ask parents to assist children in recording their findings, then go back to class and compare the sounds that were heard. Did children hear the same things? Sounds may be charted with comparisons.
- As parents invite children to help with regular cooking routines or a special dinner, children may taste, smell, feel, and observe ingredients before, during, and after the cooking process. Parents and children may make a list of everything that was experienced by each of the senses. Of course, everyone gets to eat the food and describe its texture. Making popcorn is a simple and inexpensive way for children to have a sensory experience. Making popcorn engages all of the senses. Children hear it pop, watch the transformation from corn to "puff," and smell the corn as it pops. Comparison can be made between the kernel and the finished product and between popcorn with and without salt.

Potential learning experiences for families based upon the senses present themselves every day. Teachers may want to remind parents to capitalize upon sensory experiences and heighten their awareness of the opportunities available. For example, "Taste this salt that I am adding to the stew and tell me about it." "Does the cake batter need more sugar?"

◇ Documenting and Assessing Children's Learning

Documenting and assessing children's understanding and appreciation of the five senses (and their combinations) for learning is a continuous process done on an individual and group basis using various tools and methods, including

- teacher observations and anecdotal records;
- individual and group discussions with children as they work and think by themselves and in groups;
- structured interviews with children about their sensory experiences;
- portfolios of children's work (photographs, drawings, charts, labels, artwork, correct usage of terminology, stories, feedback from home);
- examination of children's science journals;
- children's self-evaluations of their interest level and their participation in active experiences revolving around the senses;
- children's ability to formulate questions;
- children's ability to document findings;
- children's ability to communicate what they have learned to others.

Tear-Out Sheets 12-2 and 12-3 at the end of this chapter can be used at different times in the school year to chart growth in children's knowledge of and appreciation of the senses as a learning tool. Chapter 7 has a more complete discussion of documenting and assessing children's learning.

FOR THE CHILDREN

The teacher should fill the classroom with books, charts, materials, and experiments that emphasize the five senses. Younger children are not ready to associate learning through the senses with "how" the senses work, but they can understand that they acquire much new information about their environment and the things and people in it as they taste, smell, touch, see, and listen. Children will also develop many new vocabulary words to describe their sensory experiences. These should be recorded in science journals or charted by the teacher for exhibit.

The outside environment is also a wonderful place to plan activities around the senses. Children will be able to describe clouds and bugs through close observation, listen to the wind blowing the leaves or the chirp of a cricket, feel the texture of tree bark, smell flowers, and even taste berries (with the teacher's approval). They will also learn to appreciate the beauty of the outdoors. Older children may develop some strategies to preserve it. City neighborhoods are filled with opportunities for sensory experiences as well. The smells of foods cooking, the sampling of a cookie from the neighborhood bakery, the sound of traffic, the textures of old buildings, and the sight of signs and posters can all be described and recorded.

Children may also be introduced to one or more categories of sensory impairment, such as visual or hearing, and begin to understand what the impairment means for the person. They will also better understand how to assist a child who has a visual or hearing impairment in their inclusive classroom.

Begin by reading Aliki's *My Five Senses* as a general introduction. Make a big chart divided into five columns; label the columns with the five senses. Cut up some fruit and give each child a piece to explore. Ask the children to think of words that describe how the inside of the fruit looks, smells, feels, sounds when they take a bite, and tastes. Record the words on the chart. Bring in other objects for children to explore.

During an investigation on babies, children tasted various baby foods and recorded which ones they liked and disliked on a chart. The teacher then determined which were the favorites.

The following activities are clustered by the particular sense. Further activities for older children emphasize how the senses work together. All activities should be based on predictions, inquiry, and findings and the communication of findings to others. Read B. McMillan's *Sense Suspense.* Have children guess which sense is depicted on a page. (The answer is on the following page in the book.)

1. Sound

- Noise! Noise! Noise! Make sound shakers out of plastic film containers, yogurt or cottage cheese cups, or other opaque plastic containers with lids. Fill the containers partially full of dry seeds, dry beans, rice, pebbles, sand, shredded paper, coins, or marbles. Make sure the containers are well sealed. There are several activities to do with these noisemakers. First, play a guessing game. Have children guess what is inside of the sound shaker by trying it out a number of times and describing the sound. Teachers can record the answers on a chart. Next, have children compare the sounds made by different materials. Which are different? Which sound pretty much the same? Finally, children can decorate the containers, label them, and make them a permanent part of the classroom to be used for singing, dance, and movement activities. Children will also have fun at the class party having their families guess what is in the containers.

Materials:

Plastic containers

Glue or tape

Fillings (dry seeds, uncooked beans, rice, sand, pebbles, marbles, or coins)

◆ Play a CD with different sound effects. Play each sound and see whether small groups of children (listening closely) can determine what made it. Some recordings of classical music use instruments to imitate certain sounds in the environment. Have children guess what the instrument wants them to hear—sirens, the wind, a waterfall?

Materials:

Sound-effect recording or recording of classical music

Device for playing the recording,

Material for recording children's responses

◆ Read *Mandy* by Barbara Booth. Ask children what they know about hearing impairments and hearing aids. Play a video without the sound and ask children why it was difficult to understand the video. Record children's perceptions of what it would be like to have a hearing loss. If possible, have a classroom visitor who is an expert on hearing impairments demonstrate ways to compensate for a hearing loss and how children can assist their peers who are hearing-impaired. Make sure your visitor knows how to speak with young children.

Materials:

The book *Mandy*

Video

Chart

Resource person

2. Smell

◆ Collect several items that have distinctive smells. Some examples are lemon, mint, pine needles, vanilla, vinegar, onion, and cedar wood. Keep the items enclosed in plastic containers so that they do not mix. Either blindfold the children (but remember that some children are afraid of blindfolds) or punch holes in the lids and ask the children to do the following steps:

- Try to describe the odor and identify what is in the container. Record responses in journals or on chart paper.
- Rate the odor from good to bad. The rating may vary by child. Some people like smells that others avoid. Discuss why the smell is good or bad.
- Draw a picture about what the smell reminds them of. For example, vanilla may remind them of making cakes and cedar wood of camping. Have children dictate a story about their drawing.

  Materials:

  Items to smell

Children are eager to use their sense of smell.

Paper and writing implements for stories

Blindfold, or perforated container to hold "smelly" items

- Plan a walking trip outside of the classroom. The great outdoors is full of wonderful smells. Have children describe the smells and record them in the science journal or on chart paper. If you are in an area where seasons are very distinct, compare the smells of spring, summer, fall, and winter. An interesting "smelling trip" would include places around the school, such as the kitchen or cafeteria, the main office, and the library. Each has distinctive smells. Have children record the smells that they discover and compare them with the findings of other children.

  Materials:

  Science journals or chart paper

- Create "smell cards" for memory or matching type games. Children may collect and dry herbs or flowers that have a strong smell. Glue or tape the items to small pieces of cardboard.

  Materials:

  Items to smell, such as dried herbs and flowers

  Paper

  Glue or tape

3. Touch

- Gather a number of things that are cold, warm, smooth, or rough, such as sandpaper, ice, warm water, and an apple. Work in small groups. Have children close their eyes and tell you what they feel when you touch the items to their hands or fingertips.

  Materials:

  Items to touch

  Paper to record children's responses

- Have children bring items with different textures from home or from the outdoors to create a "Please Touch Me" tray. Items may be added and subtracted as children tire of them. Be sure to have the children record the item and their perception of its texture in their science journals. They can also add drawings and written labels.

  Materials:

  Large tray

  Items for tray

  Science journals and pens or markers

- Get several grades of sandpaper from a hardware store. Cut them into pieces of the same size. Be sure to write the degree of roughness on the back of each piece. This is a self-correcting activity. Mix the pieces up, have the children put them in order, and then check to see whether they are correct. If not, have them record where they made their mistake.

  Materials:

  Sandpaper of various degrees of roughness

  Science journals and writing materials

4. Taste

- Make a chart with the headings *Sweet, Bitter, Sour, Salty.* Gather several foods that fit into these four categories of taste. Cut them into small pieces. Have the children taste one at a time and vote on which category it fits into. If there is a disagreement, have the children state why they picked a category. Continue until all food items are categorized. Have children discuss which taste they prefer and why. Chart the results.

  Materials:

  Food items

  Chart paper and marker

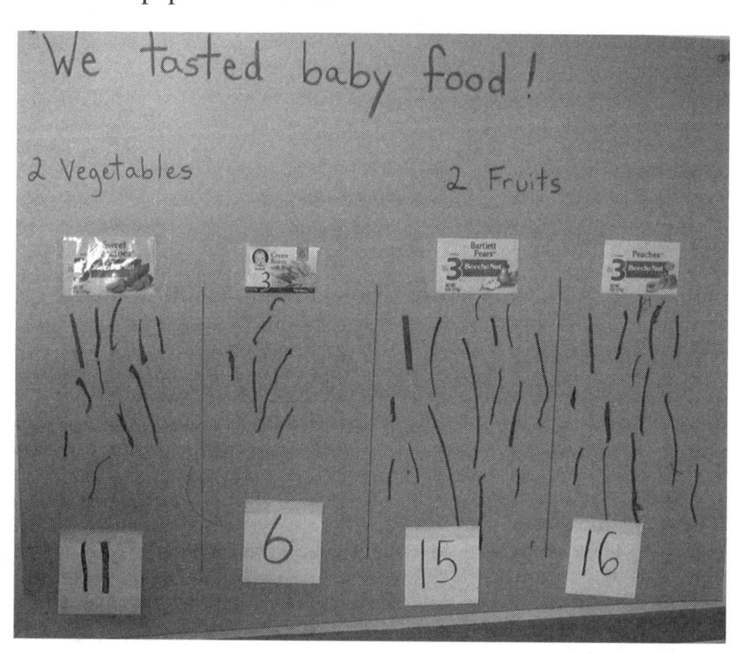

Children tasted baby food as part of an investigation on babies.

◆ Have the children decide on a cooking activity. Confine their choices to foods that require a few simple ingredients. All ingredients should be available for them to taste. Then do a taste test of the final product. Can they identify any of the individual ingredients? If so, which ones? How does the cooked product taste with all of the ingredients combined? Cakes or simple stews work well for this experience.

	Sweet	Bitter	Sour	Salty
	✓			
				✓
			✓	
	✓			
		✓		
		✓		

Materials:

Ingredients for a cooking activity of the children's choice

Cooking utensils, pots or pans

A small oven

Spoons and small plates or cups for tasting

5. Sight

Children who are not visually impaired depend on sight for most of the information that they bring in through the senses, unless they are continually reminded to use their other sensory apparatuses. Activities related to sight should be centered on developing an appreciation for information acquired visually and encouraging children to observe or look more closely at the things that they take for granted. Do things really look the way they appear at first glance? Everywhere you go with children, ask them to describe what they are seeing and encourage them to look more carefully. They may discover a small animal or bird that appears to be hidden if they look with more care.

◆ In a small group, have children attempt to describe an object while they are blindfolded or have their eyes closed. Young children will not close their eyes for long, but you can record their perceptions from touch, smell, and sound immediately. Record the vocabulary words they use. Have children guess what the object is based on their sensory observations. Then have them open their eyes and see whether they were right. Ask children how they could assist a visually impaired classmate in "visualizing" an object such as a block shape, a car, or a doll. How would they describe it? What words would they use? Have them draw a picture and dictate a story about what it would be like to be visually impaired.

◆ Set up a corner of the classroom with colored water, a prism, a microscope, and a magnifying glass. Younger children may need help with the microscope. Have them hold the prism to the light. What do they see? Observe closely. Help them record as many words as they can to describe their observations. Have them observe the colored water with the magnifying glass, then with the microscope. These scientific tools can be used with small objects found outside or with pond water to magnify their visual experiences. Have children predict what they will see when surfaces are magnified and record their predictions. Then have them describe through pictures and stories what they actually saw.

◇ Reflecting

Children can organize their experiences by using large areas of the classroom to display their experiments with the senses, their journals, the charts reflecting sensory perceptions, and their artwork. The activities previously suggested require the creation of various learning centers that children may revisit to reflect upon their work. Teachers can discuss with small groups of children the results of their sensory investigations and explorations. In fact, all of the investigations involve small group discussions to dispute, record, and communicate the results to others.

Have a party for family members to experiment with the senses. Use the centers and activities described here and have the children suggest directions to post for families so that they may also use their senses in an environment that provides several and varied interactive experiences. Because food is highly sensory, have lots of different food available for families to taste and enjoy. Both children and the cooking staff can prepare the meal. Help children make the food visually appealing by adding decorative touches. Children can ask family members if they can identify the ingredients that were part of each recipe. Give everyone a chance to admire the interesting centers and the charts of new vocabulary words. (See Tear-Out Sheet 12-1 at the end of this chapter.)

◇ Extending and Expanding to the Primary Grades

Children in the early primary grades can do the following exercises:

- Understand that the senses work together. For example, the nose is responsible for part of the flavor of food. Ask children to hold their noses closed and use just their tongues to identify what they are eating. Give them something like a pear or apple slice. They may have trouble telling you what it is. Also ask them to distinguish one food from another. Have them try again with their nose open. They will probably get it right. Baby food is an excellent vehicle for testing how the senses work together.

- Associate their senses with the parts of the body that they represent. Discuss how each of the senses functions. Show children a model of the ear, for example, and explain how it conducts sound.

- Build a larger and more complex scientific vocabulary pertaining to the senses.

- Appreciate their senses and learn how to care for them properly. Have children generate ideas on how to protect their hearing (ears) and vision (eyes), or name safety hazards to the eyes and ears. Chart these and have the children turn them into safety rules for the classroom.

- Access websites for information on the senses.

- Read *A Picture Book of Louis Braille* by David Adler. Discuss Braille's remarkable achievements and let the children close their eyes and try reading using the Braille alphabet. This will take practice. Have them write a short story using perforations instead of the usual implements.

- Read *A Picture Book of Helen Keller* by David Adler. Have the children predict what Helen will be able to do without two of her senses. Then have them imagine how they could learn using their other senses. Have them prepare short dramatizations of Helen Keller's life.

Science Exhibit and Class Party:
Everyone Come!
Child Care and Refreshments

[Date]

Dear Parents,

Our class has finished investigating the five senses. During this time, we have focused on activities and experiences designed to help your children understand how we learn through the senses. We have conducted many experiments to discover how to use our senses to understand the world around us. We hope that you enjoyed the listening and observing walks that you took with your children as well as the cooking activity.

We have organized a class party to celebrate the many things we have accomplished during this unit. We invite you and your family to our party on_______ at _______ in the _______ at school. As usual, transportation and child care will be provided for those who need it.

The event will include exhibits of all types, including your children's wonderful artwork, displays of books about the five senses, and music with distinct sounds. We will also have some fun activities for parents and children, including experiments that the children have prepared as well as simple cooking. As usual, we will have some tasty refreshments for you to enjoy.

We thank you again for working with us and look forward to seeing you at the party.

Sincerely,

Group Observation: Science Terms
(These terms refer to the five senses)

Date: ____________________________

Center/Area: ____________________

Children's Names: _______________

	Accuracy		
Science Terms Used	**Not at All**	**Some**	**Accurate**
1. smell	______	______	______
2. touch	______	______	______
3. hear	______	______	______
4. taste	______	______	______
5. sight	______	______	______
6. rough	______	______	______
7. smooth	______	______	______
8. bitter	______	______	______
9. sour	______	______	______
10. salty	______	______	______

Comments:

Individual Evaluation: Assessing Children's Science Skills
The Human Body: The Senses

Name:__________________

Date:__________________

DOB/Age:__________________

	Always	Sometimes	Never
Can name the five senses	______	______	______
Associates each sense with the correct parts of the human body	______	______	______
Uses one or more of the five senses to discover properties of things in the environment	______	______	______
Compares things using only one sense	______	______	______
Classifies things using only one sense	______	______	______
Describes how two or more senses work together	______	______	______
Appreciates the senses as a learning tool (can describe how something is being learned)	______	______	______
Demonstrates an understanding of what it is like to have sensory impairment	______	______	______
Demonstrates an understanding of persons with sensory impairment	______	______	______
Asks appropriate questions	______	______	______
Participates in sensory experiments	______	______	______
Makes appropriate predictions and conclusions	______	______	______
Uses reference books	______	______	______
Writes stories about the senses	______	______	______
Uses senses to describe indoor/outdoor environment	______	______	______

Additional Comments:

13

Healthy Bodies

FOR THE TEACHER

◇ What You'll Need to Know

Young children should have a variety of experiences that provide an initial understanding of personal health. They are anxious to grow and become strong, but teachers should be aware of the concepts and misconceptions they have about health. Most children attribute all illnesses to germs without an understanding of the different origins of disease and the prevention and cure of disease. Children do link eating with growth, health, strength, and energy, but they do not necessarily understand that some foods are nutritionally better than others—or why. Additionally, they are eager to learn about their bodies, inside and out, but they have worries about body images and differences between their development and that of others.

With the help and guidance of teachers and parents, young children need to become individuals with a lifelong responsibility for their own health and personal care, including dental hygiene, cleanliness, and exercise. Some early childhood educators have noticed that despite recommendations for good nutrition, changing family lifestyles are having an adverse impact on the nutritional habits of many young children. More sugar and fat now enter their diets, and many cooking activities suggested for classroom use do not include nutritional learning objectives. Thus, parents and teachers should be committed to the premise that the care and nourishment of the human body is essential to the well-being of children and the content of a science program. Be sure to inform parents about your emphasis on healthy bodies and enlist their help in maintaining good health practices in the home.

It is understood that both families and teachers have food allergies and preferences based on conviction and religious practice. They are free to eliminate ideas and investigations that are not consistent with their views.

◇ Key Concepts

- Each person is unique with a special body type.
- Lungs help us to breathe and use the oxygen in the air.
- Bones support our bodies and help them keep their shape.
- Our hearts are a special part of our bodies.
- We depend on muscles to move every part of the human body.
- The food pyramid (now called MyPyramid for Kids, USDA, 2005) offers many choices of good foods to keep our bodies healthy.
- We must eat foods from each of the major food groups in order to stay healthy.
- We help ourselves stay healthy and grow strong by contributing to our own personal care.
- All children have feelings.
- Children can investigate ways to express their feelings constructively.

◆ Standards Alignment

The Key Concepts and Goals and Objectives are based on the *National Science Education Standards* content standards, grades K–4. Standard F: Science in Personal and Social Perspectives (personal health).

Benchmarks for Science Literacy (The Human Organism)

◇ Goals and Objectives

The children will be able to

appreciate their uniqueness by describing their bodies and comparing themselves with others;

understand that their feelings are not unique;

examine simple diagrams of the lungs and learn how they operate through exercises using the lungs;

recognize that bone structure supports the body by examining a model or an illustration of the skeleton;

identify the heart as a muscle;

learn that bone structure and body actions are related;

identify the feeling of muscles in many parts of the body;

recognize and appreciate nourishing foods;

plan a healthy, balanced diet;

recognize the health contribution of different foods;

recognize the importance of exercise;

try new foods;

assume responsibility for washing their hands and brushing their teeth;

recognize the importance of having fun.

◇ What You'll Need

The following books are primarily for adults and older children and provide an excellent background for the early childhood teacher on the body and how it functions:

1. *Healthy Young Children: A Manual for Programs* (2002, 4th ed.) edited by S. S. Aaronson with P. M. Spahr (National Association for the Education of Young Children).
2. *Speedy Facts: You're Tall in the Morning but Shorter at Night and Other Amazing Facts about the Human Body* (2004) by M. Berger and G. Berger (Scholastic Reference).
3. *Movement-Based Learning: Academic Concepts and Physical Activity for Ages Three through Eight* (2006) by R. Clements and S. Schneider (American Alliance for Health, Physical Education, Recreation, and Dance).
4. *The Cooking Book: Fostering Young Children's Learning and Delight* (2005) by L. J. Colker (National Association for the Education of Young Children).
5. *Why Don't Your Eyelashes Grow? Curious Questions Kids Ask about the Human Body* (2008) by B. A. Ditkoff (Penguin).
6. *Eating* (1994) by A. Ganeri (Raintree).
7. *Moving* (1994) by A. Ganeri (Raintree).
8. *Bodyscope: Movers and Shapers* (2004) by P. Macnair (Kingfisher/Houghton Mifflin).
9. *My Body* (2004) by J. E. Moore (Evan Moor Publishers).
10. *I Scream, You Scream: A Feast of Food Rhymes* (1997) by L. Morrison (August House).
11. *Fit for Life* (1996) by A. Parsons (Watts).
12. *Nutrition* (1993) by D. Patent (Holiday House).

13. *Exercise and Your Health* (1998) by J. Powell (Raintree).
14. *Breathing* (1995) by A. Sandeman (Copper Beech).
15. *Guts: One Digestive System* (2005) by S. Simon (Harper Collins Children's Books).

Children's Books

Adler, D. (1991). *You Breathe In, You Breathe Out: All about Your Lungs.* Danbury, CT: Watts.

Berger, M. (1983). *Why I Cough, Sneeze, Shiver, Hiccup, and Yawn.* New York: HarperCollins.

Cole, J. (1998). *The Magic School Bus: Inside the Human Body.* New York: Putnam.

Dooley, N. (1991). *Everyone Cooks Rice.* Minneapolis: Carolrhoda Books.

Feeney, K. (2001). *Get Moving: Tips on Exercise.* Mankato, MN: Bridgestone.

Hausherr, R. (1994). *What Food Is This?* New York: Scholastic.

Hoban, R. (1986). *Bread and Jam for Frances.* New York: HarperCollins.

Hoberman, M. A. (1997). *The Seven Silly Eaters.* San Diego: Harcourt Brace.

Leedy, L. (1994). *The Edible Pyramid: Good Eating Every Day.* New York: Holiday House.

Murphy, M. (2000). *I Feel Happy and Sad and Angry and Glad.* New York: Dorling Kindersley.

Nanao, J., Hasegawa, T., & Stinchecum, A. M. (1995). *Contemplating Your Bellybutton.* San Diego: Kane/Miller Publishers.

Roca, N., & Curto, R. (2006). *The 5 Senses.* Hauppauge, NY: Barron's Educational Series.

Sandeman, A. (1995). *Bones.* Minneapolis: Copper Beech.

Showers, P. (2001). *Hear Your Heart.* New York: Harper Collins.

Showers, P. (2004). *A Drop of Blood.* New York: Harper Collins.

Sweeney, J. (2000). *Me and My Amazing Body.* New York: Dragonfly Books.

Thomas, P. (2002). *Don't Call Me Special: A First Look at Disability.* New York: Barron's Educational Series.

Thomas, P. (2002). *My Amazing Body: A First Look at Health and Fitness.* New York: Barron's Educational Series.

Wynne, P. (2009). *My First Human Body Book.* Mineola, NY: Dover Publications.

Websites

1. The USDA's *MyPyramid* website (http://www.mypyramid.gov/index.html) is an excellent resource for teachers. The site includes guidelines for giving children a healthy, balanced diet. It is full of information on children's health based on new USDA research and guidelines.
2. *MyPyramid for Preschoolers* (http://www.mypyramid.gov/preschoolers/index.html) is for children from 2 to 5 years of age. This website can be used to teach preschool children to eat well, be active, and be healthy.
3. *Head Start Body Start* (http://www.aahperd.org/headstartbodystart/) is a website dedicated to promoting physical activity, outdoor play, and healthy lifestyles for young children and their families. The website is full of resources and ideas to help bring active play and meaningful movement to early childhood programs. Parents will also find activities and tools to inspire creative, movement-based play and healthy food choices at home.

◇ Other Things You'll Need

- Chicken bones
- Chart paper
- Large mirror
- Scale
- Scissors
- Dirty fabrics
- Dishpans
- Potting soil
- Cooking utensils and cooking source
- Play props for a pretend restaurant
- Stethoscopes
- Markers and crayons
- Yardstick or tape measure
- Stiff cards and paper
- Glue
- Soap
- Water
- Plant pots and containers for planting
- Magazines for collage
- Illustrations of the human body

Learning to cook healthy food fosters healthy development.

The Home–School Connection

Today's busy families often sacrifice sleep, exercise, and a healthy balanced diet because of their multiple commitments to work and the many facets of family life. Families are partners in this area of the science curriculum because they set the stage for their children's healthy development. Every day is filled with opportunities to foster healthy development. A child's bad cold can be turned into a chance to learn science. Parents can read simple books on how diseases are passed from person to person. This knowledge sets the stage for the enforcing of simple rules such as regular teeth brushing and hand washing. These routines become hassle-free when children practice them early at home.

Or parents can teach some ways to stay healthy, such as not sharing utensils or glasses and covering our nose and mouth when we sneeze or cough. There are many fine cookbooks with simple healthy recipes that parents and children can try together. The book *Everyone Cooks Rice* by N. Dooley not only fosters a shared experience but also conveys positive images of other cultures to children. A pictorial food pyramid can hang on the refrigerator as a reminder of foods that children should sample often and others that we save for special occasions because—although they may be pleasant to eat—they don't help us grow well, strong, and healthy. The new MyPyramid for Kids says "Fats and sugars—know your limits." Parents may give children a role in planning a weekly menu.

Because of the concern about childhood obesity, in 2005 the U.S. Department of Agriculture (USDA) created new dietary guidelines to provide more practical advice on how parents can give their children a healthy, balanced diet. The new guidelines include 30 to 60 minutes of moderate to vigorous exercise daily. Parents can participate with their children for a healthier family. See the new MyPyramid.

On a regular trip to the grocery store, parents can allow children to select from the various food groups. For older children, bags may be labeled ahead of time so that children can select the foods themselves and place them into the containers. Within a food group, there are always favorites. Finally, families can encourage children to feel good about their bodies and body image. Children's growth may be charted on an unobtrusive

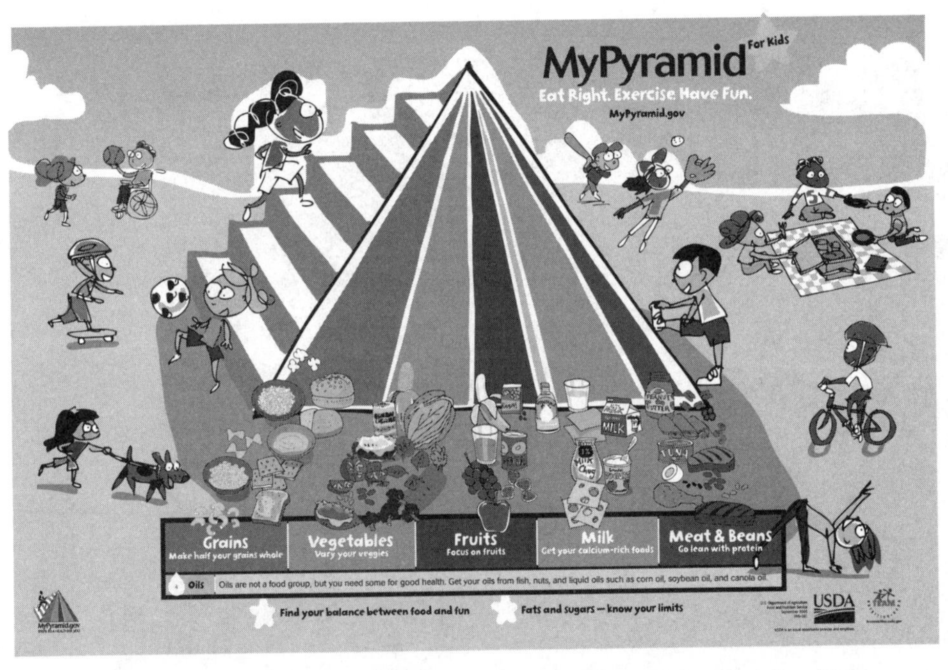

Source: U.S. Department of Agriculture, Food and Nutrition Service, "My Pyramid for Kids" (September 2005). Available at: http://www.mypyramid.gov.

wall or chart paper. Parents can and should encourage their children to view themselves and their feelings in a positive light.

◇ Documenting and Assessing Children's Learning

Documenting and assessing children's concepts of the body, how it works, and how to keep it healthy is a continuous process done on an individual and group basis using the following methods and tools:

- charts, photographs, and exhibits of all kinds to document investigations;
- teacher observations and anecdotal records;
- individual and group discussions with children as they work individually or in groups;
- structured interviews with children about the particular concept or investigation under study;
- portfolios of children's work including journals, drawings, dictated stories, and diagrams;
- children's self-evaluations of their interest level and their work.

Charts and exhibits may be thematic, such as body parts, foods we eat, staying healthy, and environmental concerns. *All About Me* books created by the children and a large chart or bulletin board with drawings and pictures of the restaurant and bakery will help children recall their new concepts about healthy bodies (see the following section "For the Children"). Encourage children to create a newsletter for parents and members of the school community to explain how to stay healthy and suggest ways to improve the school and community environments. Children can be given a role in planning menus within the food guidelines. They can also work to figure out how much exercise they need to stay in balance with their food intake.

Tear-Out Sheets 13-1 and 13-2 at the end of this chapter can be used at different times in the school year to chart growth in children's knowledge of their bodies and how to keep them healthy and to help you plan your curriculum to extend and expand upon children's current understandings.

FOR THE CHILDREN

◇ Indoor and Outdoor Activities

The teacher should emphasize the theme of healthy bodies throughout the various centers of the classroom by using displays of children's fiction and informational books, posters, photographs, diagrams, and plastic models of the body, if possible. To stimulate children's interest and concept formation, active experiences may be constructed for centers, such as a restaurant in the dramatic play area that serves a balanced and nutritious diet.

◆ Have each child make a book entitled *All About Me*. The format may vary from child to child. The children may include a self-portrait after looking in the mirror, their height, their weight, the things they like to eat, how long they like to sleep, their favorite type of exercise, and other items related to health. They may want to add pages as the year progresses to chart their growth. Each child may dictate a story explaining what makes him or her different from other class members and how they are the same.

◆ Read *You Breathe In, You Breathe Out: All About Your Lungs* by D. Adler. Point out the simple diagram explaining how we breathe. Have the children place their hands on their chest and inhale deeply. Ask if they can feel their chest get bigger as their lungs fill with air. Then have them exhale. Do their chest and lungs get smaller?

Charts, photographs, and exhibits of all kinds document investigations.

Have small groups of children exercise outside. Does their breathing get faster? The teacher can then chart the answers to these and other questions.

- To emphasize the need to keep the air that we breathe clean, have children tape up index cards at various places inside and outside the classroom. On each card, they can put a small amount of petroleum jelly. After a few days, have the children note the amount of pollutants that have built up on each card. Ask them which cards have the most and why they think so. Make a chart to document the results.

- Read selected parts of *Bones* by A. Sandeman. Remind the children that without bones to hold their organs in place, their bodies would collapse. Do a movement activity that involves collapsing and getting up again using various bones and muscles. Have children form pairs. When one child curls up tight, have the other feel small bones in the first child's spine and then switch positions. Have children dictate and illustrate a story about the bones they can feel and how they feel and bend differently.

- Acquire a plastic skeleton or make a visit to a clinic or doctor's office where children can observe and describe one.

- Take a trip to a natural history museum where the bones of various animals are exhibited. Have children prepare a list of things that they want to know. After returning to the classroom, list on chart paper the facts that children now know. Is there anything that they would still like to know? This activity may serve as the next investigation.

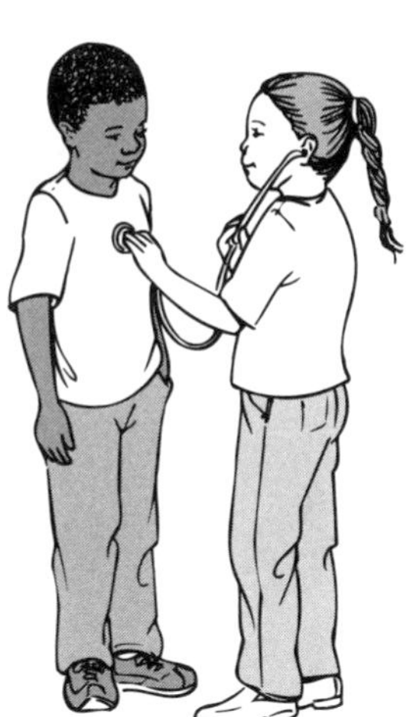

- Save chicken bones from mealtime (and have children bring them in). Put them on the science table and have children examine and describe them. How do the bones differ from their own and the others that they have observed?

- Read selections of *You Can't Make a Move Without Your Muscles* by P. Showers. Have children go to the carpeted section of the classroom or use exercise mats. Ask them what they know about muscles and record their answers. Then ask them to do a series of movements to feel the muscles working in their bodies. Movements may include moving their fingers off the floor, stretching their arms to reach the "sky," trying to touch their toes with their legs in a straight position, and curling up tightly and then unwinding. Finally, have them make different faces. Do they need muscles to smile? To frown?

- Acquire a stethoscope or visit a clinic or doctor's office. Have children form pairs and listen to each other's heart. Children can listen through paper tubes if no stethoscope is available. If the children are able to do so, count heartbeats for about 30 seconds when the children are at rest and then after exercise. Have the children dictate a group story about their observations.

- Read *The Edible Pyramid: Good Eating Every Day* by L. Leedy. The Edible Pyramid is a restaurant that serves a variety of nutritious meals. The animals at the restaurant are introduced to the food groups shown in the nutritional pyramid. Have children design a collage using old magazines. Hang a nutritional chart in the classroom. (It can be pictorial with labels.) Ask children to make sure that each food group is represented. Under each food group (grains, for example), have the children glue a picture of an example of that type of food that they like.

- Plan a visit to a good local restaurant. Speak with the owner in advance and ask the owner to explain how he or she chooses from nutritious foods to make delicious meals. Have children choose a nutritious meal, keeping the ingredients of each dish in mind.

- Providing tables and eating utensils, create a restaurant in the creative dramatic corner. Have children decide on a name and design menus, placemats, napkins, and decorations. Provide pads and pencils to take orders and make out checks. Extend this project as far as the children want to go. Create signs, plan shopping trips to the local market, cook and serve real food, and invite guests to sample the children's healthy and well-balanced meals. Foods can vary across cultures as children are exposed to various cookbooks. Be sure to respect the dietary laws and customs of different cultures. Concepts of money can be integrated by using prices on the menus and exchanging money for goods.

- Create a theme store such as a bakery. Read *Bread, Bread, Bread* by A. Morris and bake breads from different countries. Have children design a take-out menu to be distributed to small groups of parents or children at a time. Sell the bread and use the proceeds for nutritious foods.

- Plant a garden inside or outside. This, too, will be an extended project (See the investigation described in Chapter 6.) Children can cook the vegetables and grains, watch how they are transformed in the cooking process, and serve them in the restaurant or as a treat for parents and the school director. They may also create "science salads" from fruits and vegetables grown in their gardens.

- Children are very interested in germs. Find out what they know. What they may not know is that the germs that cause illness are too small to see. Have the children experiment to find out how to get those germs off their hands by washing with soap and water. In the science center, provide dirty fabric, two dishpans, soap, and water. Have children wash one piece of fabric just with water and the other with soap and water (this process can be repeated as children come to the center). Let the fabric dry and compare the two samples. Children can then record findings in their science journals.

◇ Reflecting

Ask children to organize their experiences by providing them with an area of the classroom where they can display their journals, *All About Me* books, charts, and explanations of experiments. Add menus, home-grown and home-cooked food, and props that children have created for their restaurant or bakery.

Have a party for family members to sample the foods the children have grown in the garden as well as breads and nutritionally balanced meals created for the restaurant. Parents and siblings will enjoy the fruits of their labor.

◇ Extending and Expanding to the Early Primary Grades

Children in the early primary grades can do the following activities:

- Draw a story line about a day in their lives from the time they woke up until they went to sleep at night. Include all the health and safety things they did. The teacher may write questions on poster board: When did you wash your hands? Did you brush your teeth after each meal? In the morning? At night? Did you take a shower or a bath? What situations occurred that caused you to think about safety? Children can then discuss the importance of health precautions and safety.

- Classify foods into food groups. Place foods, food replicas, or empty food cartons into bags labeled with the food groups or on a large poster or table depicting the food pyramid. They can match the food to the description of the food group. For example, *Breads and Grains—These provide energy and fiber. We need to eat the most from this group.* This would be a good game to play with family members at a party.

- Make copies of one week's menus from the school cafeteria. Discuss whether the meals meet the criteria set by the food pyramid. What about the portions?

- Learn about their teeth while making a map by carefully biting into an apple slice. Describe the different marks and learn the terms for different teeth. Finally, discuss the reasons for needing different kinds of teeth.

- Understand the reasons why the classroom should be clean and safe. In small groups or as a total group experience, children can make rules that will ensure cleanliness and safety. The rules should be recorded, hung in the classroom, and modified when necessary.

- Acquire more sophisticated knowledge of the body. Read *The Magic School Bus: Inside the Human Body* by J. Cole. Have children draw parts of the body and label them.

- Understand why it is necessary to drink water. Children can record the amount of water they drink each day and determine whether it conforms to the prescribed standards for their body.

- Check out the websites of science museums, the U.S. Department of Agriculture, and other resources for children. Teachers should make sure that the sites are suitable.

- Build a larger working vocabulary of scientific terms related to healthy bodies. These can be recorded in their science journals and posted near experiments for easy reference.

- Create a bulletin board based on the investigation of healthy bodies. Include charts for each child on how much and what type of exercise they do daily. Add up the exercise time for the whole class and try to increase it. Exercise itself merits an investigation.

- Teach children to be sensitive to environmental issues and take action to support a healthy environment. Read *The Great Kapok Tree* by L. Cherry and *Recycle! A Handbook for Kids* by G. Gibbons. The first book should help children understand the interconnectedness of elements of the environment. The second should inspire children to recycle at home, in the classroom, and in the community. Children can be encouraged to write letters to their principal, to the mayor of their town or city, or even to Congress or the president about environmental issues that concern them.

Individual Evaluation Assessing Children's Science Skills Healthy Bodies

Date: ______________

Name: ______________

DOB/Age: ______________

	Always	Sometimes	Never
Describes own body	______	______	______
Compares own body to those of other children	______	______	______
Appears to be satisfied with body image	______	______	______
Examines simple diagrams of the lungs	______	______	______
Participates in exercises using the lungs	______	______	______
Examines models or illustrations of the skeleton	______	______	______
Recognizes that bone structure supports the body	______	______	______
Identifies the heart as a muscle	______	______	______
Discovers that bone structure and body actions are related	______	______	______
Discusses the feeling of muscles in parts of the body	______	______	______
Identifies nourishing foods	______	______	______
Discusses the value of nourishing foods	______	______	______
Plans a healthy, balanced diet	______	______	______
Grows nutritious foods	______	______	______
Is willing to try new, nutritious foods	______	______	______
Assumes responsibility for washing hands	______	______	______
Assumes responsibility for brushing teeth	______	______	______
Reads books about healthy bodies	______	______	______
Consults reference books about healthy bodies	______	______	______
Writes stories about healthy bodies	______	______	______
Asks appropriate questions	______	______	______

Notes/Comments:

Group Observation: Science Terms
(These Terms Refer to Healthy Bodies)

Date: ________________

Center/Outside: ________________

Children's Names: ________________

Science Terms Used	**Accuracy**		
	Not at All	**Some**	**Accurate**
1. body type	_____	_____	_____
2. bone structure	_____	_____	_____
3. germ	_____	_____	_____
4. stethoscope	_____	_____	_____
5. exercise	_____	_____	_____
6. food group	_____	_____	_____
7. balanced diet	_____	_____	_____
8. breathing	_____	_____	_____
9. muscles	_____	_____	_____
10. food pyramid	_____	_____	_____

Notes/Comments:

14

Engaging Children with the Natural World: Environmental Education

FOR THE TEACHER

◇ What You'll Need to Know

Children's experiences in science should include interactions with the environment. By engaging with the natural world, children develop values, attitudes, comfort with, and respect for nature. In the early childhood years, environmental education should focus on the development of empathy between the child and the natural world (White & Stoecklin, 2008). Providing regular contact and positive experiences with natural environments should help children "develop their love of nature and a desire to want to protect it for their future and later generations" (White & Stoecklin, 2008, p. 5).

A recent report on *Early Childhood Environmental Education* (North American Association for Environmental Education [NAAEE], 2010) provides recommendations for developing high-quality environmental education programs for young children. According to the report, engaging children with the natural world—that is, environmental education—includes knowledge about the natural world, as well as emotions, dispositions, and skills. When they are interacting with the environment, children develop a sense of wonder and learn to appreciate the beauty and mystery of the natural world. At the same time, experiences in the natural world provide opportunities for problem solving and developing an interest and appreciation of the world around them. "Environmental education and early childhood education have common key characteristics: first-hand experiences and active participation, interdisciplinary, conceptual, process development (cognitive, affective, and behavioral) problem solving skills; and holistic approach" (Vanrony, 1999).

◇ Key Concepts

- Animals and plants have a home.
- A habitat is a place where things live and grow naturally.
- Animals and plants adapt and change.
- Living things are consumers, producers, or decomposers.
- Living and nonliving things and the environment interact.
- Living and nonliving things in the environment form ecosystems.
- There are basic necessities for life: food, water, shelter, and space.
- The necessities of life are found in the sun, air, water, and soil.

◇ Goals and Objectives

The children will be able to

develop an awareness and appreciation for the natural world;

discuss changes in the environment, including weather and seasons;

identify natural resources in the environment;

observe local environmental changes;

explore nature through interactions with plants and animals;

recognize differences between living and nonliving things;

understand that plants and animals have life cycles;

ask questions about growth and change in plants and animals;

understand that animals need things from the environment;

appreciate similarities and differences of personal characteristics among people;

enjoy the natural environment.

◆ Standards Alignment

The Key Concepts and Goals and Objectives are based on the *National Science Education Standards* content standards, grades K–4. Standards C: Life Science (the characteristics of organisms, life cycles of organisms, organisms and environments).

Benchmarks for Science Literacy (The Living Environment)

◇ What You'll Need

Consult some reference books and have them available for you and the children as you plan your experiences with the natural world. The following resources are recommended for adults and older children:

1. *Early Childhood Environmental Education: Guidelines for Excellence* (2010) written and published by NAAEE. This book identifies six key characteristics of high-quality early childhood environmental education programs.
2. *Hug a Tree: And Other Things to Do Outdoors with Young Children* (2004) by R. E. Rockwell, E. A. Sherwood, and R. A. Williams (Gryphon House). This book provides ideas for outdoor learning experiences with young children.
3. *Keepers of the Earth: Native American Stories and Environmental Activities for Children* (1991) by M. J. Caduto and J. Bruchac (Fulcrum Publishing). This book includes a selection of traditional tales from various Indian peoples. The stories are accompanied by instructions for related activities dealing with aspects of the environment.
4. *Sharing Nature with Children: A Parents' And Teachers' Nature-Awareness Guidebook* (1998) by J. B. Cornell (Ananda Publications). This guidebook is highly respected by educators worldwide. The book includes 42 nature awareness activities for children and adults.
5. *Last Child in the Woods: Saving Our Children from Nature Deficit Disorder* (2008) by R. Louv (Algonquin Books). This book created a national conversation about the disconnection between children and nature, and his message has galvanized an international movement.
6. *I Love Dirt!: 52 Activities to Help You and Your Kids Discover the Wonders of Nature* (2008) by J. Ward, S. Ghahremani, and R. Louv (Shambhala Publications). This book presents 52 open-ended activities to help engage children with the outdoors.
7. *Ecology for All Ages: Discovering Nature Through Activities for Children and Adults* (1994) by J. Hunken (Globe Pequot Press). This book includes activities and experiments to introduce ecology. It also looks at five common habitats to help improve one's observation of the natural world.
8. *Teaching Kids to Love the Earth* (1991) by Marina Lachecki (University of Minnesota Press). This publication is an excellent resource for teachers.
9. *A Child's Garden: 60 Ideas to Make Any Garden Come Alive for Children* (2004) by M. Dannenmaier (Timber). This book offers a wide range of innovative examples showing how to create special places in which children can

experience nature on their own home turf. Included are child-friendly ponds, places for pets, and private refuges.

10. *A Head Start on Science: Encouraging a Sense of Wonder* (2007) by W. C. Ritz (NSTA Press). This book contains 89 science activities that have been used by Head Start teachers.
11. *My Big World of Wonder: Activities for Learning About Nature and Using Natural Resources* (2004) by Sherri Griffin (Redleaf Press). This book includes 80 stimulating activities that will heighten children's awareness of nature.
12. *Natural Wonders: A Guide to Early Childhood for Environmental Educators* (2002) by Marcie Minnesota and Marcie Oltman (Early Childhood Consortium). This guide is available as a free PDF at http://www.seek.state.mn.us. It is an excellent resource for teachers who want to engage children with the natural environment.

Children's Books

Anthony, J. (1999). *In a Nutshell.* Nevada City, CA: Dawn Publications.

Cherry, L. (2000). *The Great Kapok Tree: A Tale of the Amazon Forest.* Orlando, FL: Houghton Mifflin Harcourt.

Elhert, L. (1991). *Red Leaf, Yellow Leaf.* New York: Harcourt Brace.

Fleming, D. (1993). *In the Small, Small Pond.* New York: Henry Holt.

Fredericks, A. D. (2005). *Near One Cattail: Turtles, Logs, and Leaping Frogs.* Nevada City: CA: Dawn Publications.

Fredricks, A. D., & Dirubbio, J. (2002). *In One Tidepool: Crabs, Snails, and Salty Tails.* Nevada City, CA: Dawn Publications.

Hall, Z. (1996). *The Apple Pie Tree.* New York: Blue Sky Press.

Iverson, D. (2003). *Discover the Seasons.* Nevada City, CA: Dawn Publications.

Iverson, D. (1998). *My Favorite Tree: Terrific Trees of North America.* Nevada City, CA: Dawn Publications.

Miller, D. S. (2002). *Are Trees Alive?* New York: Walker Publishing Company.

O'Halloran, S. (1995). *The Hunt for Spring.* Yorktown, VA: Riverbank Press.

Pratt-Serafini, K. J. (2004). *Salamander Rain: A Lake and Pond Journal.* Nevada City, CA: Dawn Publications.

Robbins, K. (1998). *Autumn Leaves.* New York: Scholastic Press.

Schimmel, S. (1994). *Dear Children of the Earth: A Letter Home.* Chanhassen, MN: Northword.

Schimmel, S. (2002). *Children of the Earth: Remember.* Chanhassen, MN: Northword.

Steinberg, L. (2005). *All Around Me I See.* Nevada City, CA: Dawn Publications.

Silverstein, S. (1999). *The Giving Tree.* New York: Harper Collins.

Ward, J. (2005). *Forest Bright, Forest Night.* Nevada City, CA: Dawn Publications.

Watt, M. (2008). *Scaredy Squirrel.* Tonawanda, NY: Kids Can Press.

Websites

1. *Growing Up Wild: Exploring Nature with Young Children* (http://www.projectwild.org/growingupwild/about.htm) is the Project Wild website for early childhood educators. The guide includes 27 activities to help connect children to nature.

2. The U. S Environmental Protection Agency has a website for its *Environmental Education* division (http://www.epa.gov/enviroed/basic.html). The website includes a wealth of resources for environmental education.
3. *The Children and Nature Network* (C&NN) website (http://www.childrenand-nature.org) was created to encourage and support those who want to reconnect children with nature.

◇ Other Things You'll Need

- Large and small magnifiers
- Sieves
- Bug collectors
- Child-sized gardening tools
- Hoops
- Large cloth or sheet
- Brown bags
- Collection trays
- Photographs of plants and animals
- Nets
- Yardstick
- String
- Small cotton bags

The Home–School Connection

There are many ways in which parents can help engage children with the natural environment. Perhaps the most important is to provide opportunities for children to play outdoors and to directly experience the natural world. We know that the way children experience the natural environment differs from the way adults see nature. Instead of seeing nature as background for events, children see nature as a stimulator for their activities. In other words, children "judge nature not by its aesthetics, but rather by the nature of their interactions with it" (White & Stoecklin, 2008, p. 1). For this reason, parents should provide children with regular experiences with nature, and they should model enjoyment and respect for nature.

Outdoor play provides an opportunity for children to value nature. They will then see nature as an important part of our world. The National Wildlife Federation (NWF) (http://www.nwf.org) recommends that parents make a daily "green hour" a new family tradition: a time for unstructured play and interaction with the natural world. Check out the NWF website (http://www.nwf.org/Get-Outside/Be-Out-There/Activities.aspx) for dozens of green hour ideas.

Parents can help children find nature trails and provide opportunities for them to play as opposed to simply hiking the trail. Choose an easy route for young children and be prepared to stop frequently. Children can look for interesting things on the trail, such as bugs, sticks, pebbles, or leaves. Parents and children can invent their own nature games. For example, "Let's see who can be the first to find five different leaves" or "Let's look for animal tracks on the ground." Help children look for critters, animal footprints, and animal holes or dens and listen to the sound of the birds.

The back yard can be an exciting place for young children. Parents can help by allowing children to have a small patch of land as "their own." Let children use their patch of land in their own imaginative way, such as a building, digging, or perhaps planting. Place an old towel or cardboard on bare dirt. In two or three days, lift the towel to see how many critters have found shelter there. Let children observe these creatures. Help them count the number of different small animals under the towel.

◇ Documenting and Assessing Children's Learning

Assessing what children learn as a result of their interactions with the natural world is a continuous process done on both an individual and group basis. As described in the previous chapters, this assessment includes observations, discussions, interviews, checklists, portfolios, and self-evaluations. The tear-out sheets at the end of this chapter allow teachers to chart growth in children's knowledge of science concepts and help them plan a curriculum on the basis of children's current understanding.

FOR THE CHILDREN

Although most of the experiences will be outdoors in the natural environment, some follow-up activities will occur in the classroom. Look around your school and try to find interesting places for your children to explore. Even the plainest-looking playground can be adapted to provide exciting experiences for young children. Remember that for young children, the outdoor environment is an exciting place in which their skills of observing, comparing, classifying, questioning, and discovering can be developed. You should begin with simple experiences because young children learn best through experiences that relate to what is already familiar and comfortable. Thus, the best place to start is in an environment that is similar to what they already know, such as the school yard or a city park within walking distance.

◆ Have the children go on a schoolyard nature scavenger hunt. Provide specific directions to help develop children's observation skills and discover interrelationships in the natural world. Have the children find

- something red;
- evidence of an animal;
- two different leaves;
- two seeds;
- a small twig;
- something smooth;
- something rough;
- something prickly;
- something interesting;
- something small.

Children love to explore trees.

The activity can be modified if access is available to different habitats, such as woodland, grassland, or seashore. The scavenger hunt can also be modified; examples include a critter scavenger hunt (looking for an ant, a cricket, worm, or other small animals) or a wood scavenger hunt (looking for piece of bark, leaf, twig, smallest tree, largest tree, and so on). Decide up front whether children will have a checklist, draw pictures in their science log, or actually bring the items back. Have everyone return to a common area to share stories about their discoveries. Display the items in a classroom-based Nature Museum.

- Help the children create a schoolyard habitat. Start with a small area. There is no minimum amount of land needed for a schoolyard habitat. You might want to simply build planter boxes or add a bird feeder and bird bath to an area of the asphalt playground. You also might want to focus on attracting and supporting local insects or butterflies. First, have the students brainstorm and discuss potential sites for the schoolyard habitat. Remind the children that in order to survive, all animals need food, water, shelter, and a safe place to raise their young. Teachers should also research habitat types and determine what to plant. The children can plant seeds and help care for their new habitat.

- Have students explore a habitat by placing a hula hoop on the ground and then counting and recording the number of different plants and small animals they find within it. Compare two different habitats, such as woodland and grassland. Have students draw pictures of what they see. In the classroom, help students create a graph of the numbers of different plants or animals they found.

For young children, the outdoor environment is an exciting place.

- Adopting a tree is an excellent activity for young children. Find a tree that's within walking distance to the classroom. Have the students first use string to measure the circumference of the trunk. Then, have them use paper and crayon to make bark rubbings. Take a large piece of cloth, such as a painters' drop cloth or an old sheet, and lay it on the ground under a large branch. Shake the tree or branch and have the children observe all the things that fall on to the cloth. Look for small animals, dead leaves, small twigs, and so on. Observe the tree at different times of the year as the seasons change from spring to fall.

- Explore leaves in the schoolyard. Have the students find as many different leaves as they can. Use crayons to make rubbings of the leaves. Let the students search for leaves of different shades and colors and create a "leaf rainbow."

- Have students build a class compost bin. Construct a simple compost bin or just use an old plastic container or bin. Composting is a great way to enhance the richness of soil. Locate the bin in a shady, well-drained area. The bin should either be placed on top of wooden slats, or the bottom should be made of wooden slats, leaving spaces for air to flow through the pile. The micro- and macroorganisms living and working in the compost pile need air to survive and work efficiently. Add layers of brown (e.g., leaves, straw, shredded newspaper) and green (e.g., fresh yard clippings, fruit, vegetables). Make sure that the material is moist. The children can take turns adding material to the compost. The children can examine the compost bin from time to time and observe as the materials break down into a crumbly soil texture.

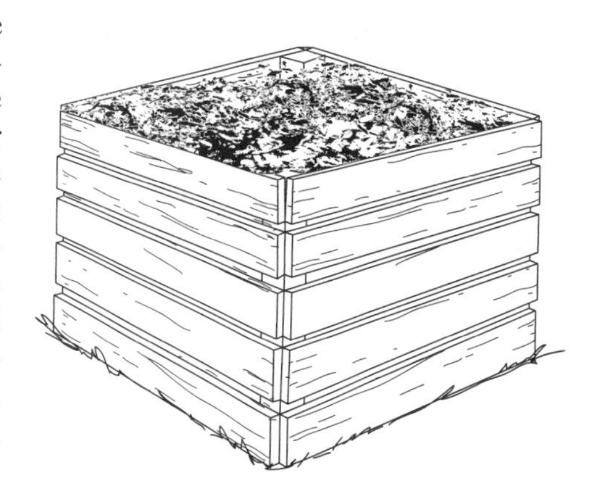

- Hiking along a yardstick is an excellent way for young children to explore nature. For this activity, children can work in pairs. Each pair is given a yardstick, which they lay on the ground. The task is to carefully explore the area along the yardstick and look for signs of insects, birds, or animals. They can also look for other distinctive signs along the trails, such as the texture of the soil or sand.

- Build a worm bin with students. This easy project is another great way to compost in a school setting because it can be done inside in compact containers. It is also a great way to show students how to provide a habitat for creatures. Teachers must provide the proper food, water, and shelter that the worms and microorganisms need in order to survive. Construct a worm bin that is about 10 to 18 inches deep and make sure that it has a tight-fitting lid and holes in the bottom or side for ventilation. Place strips of newspaper or shredded cardboard in the bin, filling about halfway. Moisten the bed and add two or three handfuls of soil. Finally, add the worms. You can purchase worms from your local bait shop. Add kitchen scraps for food and place the bin where it will not freeze or get too warm.

- You can use the soil from your compost or from your worm bin to fill a container garden. All types of containers can be used for growing plants, such as terracotta pots, plastic tubs, plastic buckets, or a wooden window box. Crops such as tomatoes, lettuce, radish, and peppers will easily grow in containers. Children can compare the growth of plants in containers with and without compost.

- Establish a pile of logs in area of the school yard that is not frequently used. Ideally, select some large logs that are already starting to rot. Explain to the children that the purpose of the log pile is to provide food and shelter to a wide range of small invertebrates. Find a damp, shady location and work together to build the log pile. Have the children check the pile regularly to see what they can find.

◇ Reflecting

Help the children document what they have learned. Provide them with an area of the classroom to display their drawings. Have the children use digital cameras to take photographs of their outdoor project, including elements such as their tree, garden, or pile of logs. Display the photographs in the classroom. Children should also have an area in the class where they can display all of the things they have collected in the school yard, log pile, or from their tree. Provide children with labels for the items they collected.

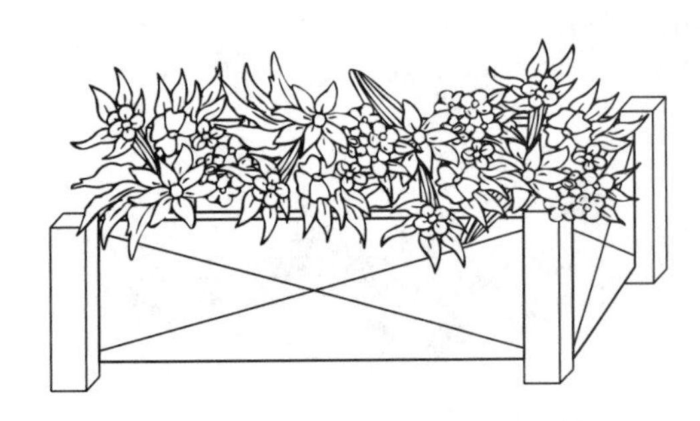

Arrange an outdoor party or picnic for family members. The food might be the crops from your own container garden. Have the children guide their family members through your schoolyard habitat, observe your adopted tree, and study the pile of logs. Use paper products for your picnic so that they can be recycled in the compost bin.

◇ Extending and Expanding to the Early Primary Grades

Children in the early primary grades can do the following activities:

- Fun with leaf litter. Leaf litter is dead plant material—such as leaves, bark, seeds, and twigs—that has fallen to the ground. Litter provides a habitat for small animals, fungi, and plants. As litter decomposes, nutrients are released to the environment. Young children love to explore the animals found in leaf litter. Introduce and read *Leaf Litter: Exploring the Mysteries of a Hidden World* by Rachel Tonkin. Put a pile of leaf litter on an old plastic tray or a piece of white paper. Provide students with insect aspirators, collection jars or bug boxes, magnifying glasses, and a sieve.

Show children how to collect insects and help them identify the insects. Release all insects after review. The sieves can be used to sift the leaf litter.

- Discover the differences in the communities of plants and animals living in two different microhabitats. Find a site on the school grounds that contains two extremely different habitats (e.g., a sunny, dry area and a shady, moist area). Lead the class to the study site and count the numbers of different plants and animals that they find in each microhabitat. Create a graph to display the findings.

- Introduce students to a more complex vocabulary, including words such as habitat, ecosystem, diversity, species.

- Have the students access websites for information on the environment.

- Teach students about ant activity. Look around the school grounds and locate a sunny spot containing ant mounds. Observe the ants' home and have the children describe what they see. Have the students make several small peanut-butter-and-jelly sandwiches (no more than 2 inches square). Have the students place one sandwich near each mound. Observe the feeding activity of the ants for 10 to 15 minutes. Take a photograph of the sandwich every 5 minutes. Have the children record their observations in their science journals.

- Help the children find different tree seeds, such as acorns or beech nuts. Plant the seeds in compost and observe the growth of the young tree. Choose a site in the school yard for your new tree and plant the tree. Measure the growth of the tree every 2 or 3 weeks.

Group Observation: Science Terms
(These terms refer to the environment and exploring nature.)

Date: ___________________

Center/Area: ___________________

Children's Names: ___________________

Science Terms Used	Accuracy: Not at All	Some	Accurate
1. nature	______	______	______
2. habitat	______	______	______
3. ecosystem	______	______	______
4. wildlife	______	______	______
5. environment	______	______	______
6. compost	______	______	______
7. decompose	______	______	______
8. shelter	______	______	______
9. plants	______	______	______
10. animals	______	______	______

Notes/Comments:

Individual Evaluation: Assessing Children's Science Skills
The Environment

Name:__________________

Date:__________________

DOB/Age:__________________

	Always	Sometimes	Never
Observes and discusses changes in the environment	______	______	______
Uses observations to make predictions about the environment	______	______	______
Demonstrates increasing ability to predict possible outcomes as a result of environmental explorations	______	______	______
Expresses an increasing appreciation and affinity for nature	______	______	______
Demonstrates a sense of respect for nature	______	______	______
Demonstrates an understanding of the importance of natural resources	______	______	______
Recognizes differences between living and nonliving things	______	______	______
Observes changes in living things over time	______	______	______
Understands that plants and animals have life cycles	______	______	______
Asks questions about growth and changes in plants and animals	______	______	______
Asks appropriate questions	______	______	______
Makes appropriate predictions and conclusions	______	______	______
Uses reference books	______	______	______
Writes stories about nature	______	______	______
Uses senses to describe indoor/outdoor environment	______	______	______

Notes/Comments

References

Aaronson, S. S., (Ed.), with P. M. Spahr. (2002). *Healthy young children: A manual for programs* (4th ed.). Washington, DC: National Association for the Education of Young Children.

Agler, L. (1990). *Liquid explorations.* Berkeley, CA: Lawrence Hall of Science.

American Association for the Advancement of Science. (1989). *Science for all Americans.* New York: Oxford University Press.

American Association for the Advancement of Science. (1993). *Benchmarks for science literacy.* New York: Oxford University Press.

American Association for the Advancement of Science. (1998). *Dialogue on early childhood science, mathematics, and technology education.* Washington, DC: Author.

Barbour, N., Barbour, C., & Scully, P. A. (2010). *Families, schools, and communities: Building partnerships for educating children.* Upper Saddle River, NJ: Merrill.

Barclay, K., Benelli, C., & Schoon, S. (1999). Making the connection! Science and literature. *Childhood Education, 75,* 146–152.

Barclay, K., & Traser, L. (1999). Supporting young researchers as they write to learn. *Childhood Education, 75,* 215–224.

Berk, L. E., & Winsler, A. (1995). *Scaffolding children's learning: Vygotsky and early childhood education.* Washington, DC: National Association for the Education of Young Children.

Bredekamp, S. (1993). Reflections on Reggio Emilia. *Young Children, 13,* 14–19.

Bredekamp, S., & Copple, C. (Eds.). (1997). *Developmentally appropriate practice in early childhood programs* (Rev. ed.). Washington, DC: National Association for the Education of Young Children.

Bredekamp, S., & Rosegrant, T. (1995). *Reaching potentials: Transforming early childhood curriculum and assessment* (Vol. 2). Washington, DC: National Association for the Education of Young Children.

Brewer, J. A. (1998). *Introduction to early childhood education: Preschool through primary grades* (3rd ed.). Boston: Allyn & Bacon.

Bronfenbrenner, U. (1979). *The ecology of human development: Experiments by nature and design.* Cambridge, MA: Harvard University Press.

Bronson, M. B. (1995). *The right stuff.* Washington, DC: National Association for the Education of Young Children.

Brunton P., & Thornton L. (2010). *Science in the early years: Building firm foundations from birth to five*. London: Sage.

Burnie, D. (2004). *Endangered planet.* Boston: Kingfisher.

Caldwell, L. B. (2002). *The Reggio approach to early childhood education: Bringing learning to life.* New York: Teachers College Press.

Carin, A. A., Bass, J. E., & Contant, T. L. (2004). *Teaching Science as Inquiry* (10th ed.). Upper Saddle River, NJ: Merrill, Prentice-Hall.

Ceppi, G., & Zini, C. (1998). *Children, spaces, relations: Metaproject for an environment for young children.* Report: ED472379.

Chalufour, I., & Worth, K. (2004). *Building structures with young children.* St. Paul, MN: Redleaf Press.

Chalufour, I., & Worth, K. (2005). *Exploring water with young children.* St. Paul, MN: Redleaf Press.

Chrisman, K. (2005). The nuts and bolts of discovery centers. *Science and Children, 43*(3), 21–23.

Clements, R., & Schneider, S. (2006). *Movement-based learning: Academic concepts and physical activity for ages three through eight.* Reston, VA: American Alliance for Health, Physical Education, Recreation, and Dance. (Available through the National Association for the Education of Young Children.)

Committee on Development of an Addendum to the National Science Education Standards on Scientific Inquiry. (2000). *Inquiry and the national science education standards: A guide for teaching and learning.* Washington, DC: National Academy Press.

Copple, C., & Bredekamp, S. (2006). *Basics of developmentally appropriate practice.* Washington, DC: National Association for the Education of Young Children.

Davis, J. (2001). The day Pork Chop died (almost). Caregivers Corner. *Young Children, 56*(3), 85.

Dewey, J. (1938). *Experience and education.* New York: Collier Books.

Dewey, J. (1944). *Democracy and education.* New York: The Free Press.

Dighe, J., Calomiris, Z., & Van Zutphen, C. (1998). Nurturing the language of art in children. *Young Children, 53*(1), 4–9.

Duschl, R. A., Schweingruber, H. A., & Shouse, A. W. (2007). *Taking science to school: learning and teaching science in grades K–8.* Washington, DC: National Academy Press.

Epstein, J. L. (1991). *Effects on student achievement of teachers' practices of parent involvement.* Paper presented at the annual meeting of the American Educational Research Association, Chicago.

Fenichel, M., & Schweingruber, H. A. (2010). *Surrounded by science: Learning science in informal environments.* Washington, DC: National Academy Press.

Fleer, M. (2008). *Everyday learning about how things work.* Canberra, Australia: Early Childhood Australia.

Gelman, R., & Brenneman, K. (2004). Science learning pathways for young children. *Early Childhood Research Quarterly, 19,* 150–158.

Goodison, T. A. (2002). Learning with ICT at primary level: Pupils' perceptions. *Journal of Computer Assisted Learning, 18,* 282–295.

Gould, G., & Sullivan, J. (1999). *The inclusive early childhood classroom: Easy ways to adapt learning centers for all children.* Beltsville, MD: Gryphon.

Grolnick, W. S., & Slowiaczek, M. L. (1994). Parents' involvement in children's schooling: A multi-dimensional conceptualization and motivational model. *Child Development, 65,* 237–252.

Harlan, J. D., & Rivkin. M. S. (2004). *Science experiences for the early childhood years: An integrated affective approach* (8th ed.). Upper Saddle River, NJ: Merrill/Prentice Hall.

Harlen, W. (2000). *Teaching of science in primary schools.* London: Fulton.

Helm, J. H., & Katz, L. G. (2001). *Young investigators: The project approach in the early years.* New York: Teachers College Press and Washington, DC: National Association for the Education of Young Children.

Isbell, R., & Exelby, B. (2001). *Early learning environments that work.* Beltsville, MD: Gryphon.

Isbell, C., & Isbell, R. (2005). *The inclusive learning center book for preschool children with special needs.* Beltsville, MD: Gryphon.

Jones, J., & Courtney, R. (2002). Documenting early science learning. *Young Children, 50*(5), 34–40.

Jones, I., Lake, V. E., & Lin, M. (2008). Early childhood science process skills: Social and developmental considerations. In Spodek, B., & Saracho, O. (Eds.), *Mathematics, Science and Technology in Early Childhood Education* (pp. 17–39). Charlotte, NC: Information Age Publishing.

Katz, L. (1993). What can we learn from Reggio Emilia? In C. Edwards, L. Gandini, & G. Forman (Eds.), *The hundred languages of children* (pp. 19–41). Norwood, NJ: Ablex.

Kilmer, S. J., & Hofman, H. (1995). In S. Bredekamp & T. Rosegrant (Eds.), *Reaching potentials: Transforming early childhood curriculum and assessment* (Vol. 2, pp. 43–63). Washington, DC: National Association for the Education of Young Children.

Klein, M. D., Cook, R. E., & Richardson-Gibbs, A. M. (2001). *Strategies for including children with special needs in early childhood settings.* Albany, NY: Delmar.

Kneidel, S. (1993). *Creepy crawlies and the scientific method: More than 100 hands-on science experiments for children.* Golden, CO: Fulcrum.

Lake, V. E., & Jones, I. (in press). *Service learning in the PreK-3 classroom: The what, why, and how-to guide for every teacher.* Minneapolis, MN: Free Spirit.

Lowery, L. (Ed.). (1997). *NSTA pathways to the science standard: Elementary school edition.* Arlington, VA: National Science Teachers Association.

Mallory, B. L. (1998). Educating young children with developmental differences: Principles of inclusive practice. In C. Seefeldt & A. Galper (Eds.), *Continuing issues in early childhood education* (pp. 214–232). Upper Saddle River, NJ: Merrill/Prentice Hall.

Mathematical Sciences Education Board on Science Education. (2005). *Mathematical and scientific development in early childhood: A workshop summary.* Washington, DC: National Academy Press.

McAfee, O., Leong, D. J., and Bodrova, E. (2004). *Basics of assessment: A primer for early childhood educators.* Washington, DC: National Association for the Education of Young Children.

McNair, S. (2006). *Start young: Early childhood science activities.* Washington, DC: National Science Teachers Association.

Michaels, S., Shouse, A. W., & Schweingruber, H. A. (2007). *Ready, set, science! Putting research to practice in K–8 science classrooms.* Washington, DC: National Academy Press.

Moriarty, R. F. (2002). Helping teachers develop as facilitators of three- to five-year-olds' science inquiry. *Young Children, 57*(5), 20–24.

Moore, J. E. J. (2000). *Learning to be a scientist.* Monterey, CA: *Evan-Moore.*

Morrison, G. (2004). *Nature in the neighborhood.* New York: Houghton Mifflin/Walter Lorraine Books.

National Academy of Sciences. (1995). *National science education standards.* Washington, DC: Author.

National Research Council. (1996). *National science education standards.* Washington, DC: National Academy Press.

North American Association for the Advancement of Environmental Education. (2010). Early childhood environmental education. Washington, DC: Author.

Paulu, N. (1992). *Helping your child learn science.* Washington, DC: U.S. Department of Education, Office of Educational Research and Improvement.

Piaget, J. (1954). *The construction of reality in the child.* New York: Basic Books.

Piaget, J. (1973). *The children's conception of the world.* St. Albans Hertfordshire, England: Paladin.

Rich, S. (2010). *Outdoor science: A practical guide.* Washington, DC: National Science Teachers Association Press.

Ritz, W. C. (2007). *A Head Start on Science: Encouraging a Sense of Wonder*. Washington, DC: National Science Teachers Association Press.

Rivkin, M. S. (1995). *The great outdoors: Restoring children's right to play outside.* Washington, DC: National Association for the Education of Young Children.

Seefeldt, C. (1995). Art: A serious work. *Young Children, 50*(3), 39–54.

Seefeldt, C. (1997). *Social studies for the preschool-primary child.* Upper Saddle River, NJ: Merrill/Prentice Hall.

Seefeldt, C. (2005). *How to work with standards in the early childhood classroom.* New York: Teachers College Press.

Seefeldt, C., & Barbour, N. (1998). *Early childhood education: An introduction.* Upper Saddle River, NJ: Merrill/Prentice Hall.

Seefeldt, C., & Galper, A. (2005). *Active experiences for active children: Social studies* (2nd ed.). Upper Saddle River, NJ: Merrill/Prentice Hall.

Seefeldt, C., & Wasik, B. (2002). *Kindergarten: Fours and fives go to school*. Upper Saddle River, NJ: Merrill.

Shonkoff, J. P., & Phillips, D. A. (2000). *From neurons to neighborhoods: The science of early childhood development.* Washington, DC: National Academy Press.

Torquati, J., & Barber, J. (2005). Dancing with trees: Infants and toddlers in the garden. *Young Children, 60,* 40–46.

Trundle, K. C., & Troland, T. H. (2005). The moon in children's literature. *Science & Children, 43,* 40–43.

Vanrony, M. (1999). Nurturing nature, environmental education for young children. Saint Paul: MN: Minnesota Children's Museum.

Vygotsky, L. (1978). *Thought and language.* Cambridge, MA: MIT Press.

Vygotsky, L. (1986). *Thought and language* (Rev. ed.). Cambridge, MA: MIT Press.

Wadsworth, G. (2003). *Benjamin Banneker: Pioneering Scientist.* Minneapolis, MN: Carolrhoda Books.

Wenner, G. (1993). Relationship between science knowledge levels and beliefs toward science instruction held by preservice elementary teachers. *Journal of Science Education and Technology, 2,* 461–468.

White, J. (2008). *Playing and learning outdoors: Making provision for high-quality experiences in the outdoor environment.* New York: Routledge.

White, R., & Stoecklin, V. L. (2008). Nurturing children's biophilia: Developmentally appropriate environmental education for young children. Retrieved from: http://www.whitehutchinson.com/children/articles/nurturing.shtml

Wick, W. (1997). *A drop of water: A book of science and wonder.* New York: Scholastic Trade.

Wilson, R. A. (1995). Nature and young children: A natural connection. *Young Children, 50*(6), 4–7.

Worth, K., & Grollman, S. (2003). *Worms, shadows, and whirlpools: Science in the early childhood classroom.* Washington, DC: National Association for the Education of Young Children.

Woyke, P. P. (2004). Hopping frogs and trail walks. *Young Children, 59*(1), 82–85.

Wright, J. L., & Shade, D. D. (1994). *Young children: Active learners in a technological age.* Washington, DC: National Association for the Education of Young Children.

Youniss, J., & Damon, W. (1992). Social construction of Piaget's theory. In H. Beilin & P. B. Pufall (Eds.), *Piaget's theory: Prospects and possibilities* (pp. 267–286). Hillsdale, NJ: Erlbaum.

Index

E

F

G

H

I

J

K

L

U

V

W

Y

Z